T0210673

Urban Art and the City

This book offers original interdisciplinary insights into cities as a diachronic creation of urban art. It engages in a sequence of historical perspectives to examine urban space as an object of apparent quasi-cycles and processes of constitution, exaltation, imitation, contestation, and redemption through art.

Urban art transforms the city into a human-made sublime which is explored in the context of the Eastern Mediterranean. The book probes this process primarily through the example of Athens and Byzantine Constantinople, but also Jerusalem, Cyprus, and regional cities, revealing how urban space unavoidably encompasses a spatial and temporal palimpsest which is constantly emerging. It presents new ideas for both the theorization and sensuous conception of artistic reality, architecture, and planning attributes. These extend from archaic, classical, and Byzantine urban splendour to current urban decline as constitution and attack on the sublime and back. Urban processes of contestation and redemption respond recently to the new 'imperialism of debt' and the positivist, technocratic understandings and demands of Euro-governments and neoliberal institutions, while still evoking older forms of spatial power.

Offering fresh notions on art, architecture, space, antiquity, (post)-modernity, and politics of the region, this book will appeal to scholars and students of geography, urban studies, art, restoration, and film theory, architecture, landscape design, planning, anthropology, sociology, and history.

Argyro Loukaki is Professor of Greek Art, Architecture, and Urban Planning at the Hellenic Open University (HOU). Her research interests include space conception, representation, and aesthetics; art in the urban and architectonic space; cultural heritage and restoration of monuments; Mediterranean cultural geography, art, architecture, and landscape; the geographical unconscious; and links between architecture, art, planning, and literature. Her previously published titles with Routledge are *Living Ruins, Value Conflicts* and *The Geographical Unconscious*.

Routledge Critical Studies in Urbanism and the City

This series offers a forum for cutting-edge and original research that explores different aspects of the city. Titles within this series critically engage with, question and challenge contemporary theory and concepts to extend current debates and pave the way for new critical perspectives on the city. This series explores a range of social, political, economic, cultural and spatial concepts, offering innovative and vibrant contributions, international perspectives and interdisciplinary engagements with the city from across the social sciences and humanities.

Socially Engaged Art and the Neoliberal City
Cecilie Sachs Olsen

Peri-Urban China
Land Use, Growth, and Integrated Urban – Rural Development
Li Tian and Yan Guo

Spatial Complexity in Urban Design Research
Graph Visualization Tools for Communities and their Contexts
Jamie O'Brien

Urban Neighbourhood Formations
Boundaries, Narrations and Intimacies
Edited by Hilal Alkan and Nazan Maksudyan

Urban Ethics
Conflicts Over the Good and Proper Life in Cities
Edited by Moritz Ege and Johannes Moser

Urban Art and the City
Creating, Destroying, and Reclaiming the Sublime
Edited by Argyro Loukaki

For more information about this series, please visit www.routledge.com/Routledge-Critical-Studies-in-Urbanism-and-the-City/book-series/RSCUC

Urban Art and the City

Creating, Destroying, and Reclaiming
the Sublime

Edited by Argyro Loukaki

Routledge
Taylor & Francis Group

LONDON AND NEW YORK

First published 2021
by Routledge
2 Park Square, Milton Park, Abingdon, Oxon OX14 4RN

and by Routledge
605 Third Avenue, New York, NY 10017

First issued in paperback 2022

Routledge is an imprint of the Taylor & Francis Group, an informa business

British Library Cataloguing-in-Publication Data
A catalogue record for this book is available from the British Library

Library of Congress Cataloging-in-Publication Data
A catalog record for this book has been requested

ISBN: 978-0-367-13296-5 (hbk)
ISBN: 978-0-367-54698-4 (pbk)
ISBN: 978-0-367-13297-2 (ebk)

DOI: 10.4324/9780367132972

Typeset in Times New Roman
by Apex CoVantage, LLC

Contents

Notes on contributors

Jenny Albani is an architect and art historian at the Hellenic Ministry of Culture and Sports, and also teaches at the Hellenic Open University. She studied architecture at the National Technical University, Athens, and art history at the Vienna University. Her research interests include the Byzantine and post-Byzantine art and architecture history. Albani has participated in excavations in Greece and Egypt, published and edited books, written articles for academic journals and edited volumes, participated in international conferences, and curated archaeological exhibitions.

Stavros Alifragkis holds a diploma from the Department of Architecture, Aristotle University of Thessaloniki; an MPhil from the Department of Architecture, University of Cambridge; an MPhil from the School of Architecture, NTU Athens; and a PhD from the University of Cambridge. His doctoral thesis focuses on the cinematic reconstruction of the Socialist city of the future in Vertov's 'Man with the Movie Camera.' Postdoctoral research was conducted at the Department of Architecture, University of Thessaly. He has taught undergraduate courses at the Hellenic Open University, the University of Patras, the Hellenic Army Academy, the University of Ioannina, and the Aristotle University of Thessaloniki. He has participated in conferences with papers on cinematic cities and film festivals with linear/interactive moving-image projects.

Charalampos G. Chotzakoglou studied history, archaeology, and the history of art in Athens, Tubing, and Vienna. He has participated in research projects of the Austrian Academy of Sciences, National Hellenic Research Foundation, Hungarian Eötvös Lorand University, and European Institute of Budapest. He is a teaching associate at the Hellenic Open University and was visiting professor at the University of Cyprus, Neapolis and Lefkosia. His book on the destruction of the religious monuments in the Turkish-occupied northern part of Cyprus was honorary presented to the Plenary Session of the Academy of Athens. He has been awarded by the Polish Orthodox Church with the medal of 'Prince Constantin Ostrogsky.'

Vicky Foskolou studied history and archeology at the University of Athens and continued her postgraduate studies at the University of Heidelberg. She received her PhD from the University of Athens. She is assistant professor in Byzantine Archeology at the Department of History and Archeology of the University of Crete (see www.history-archaeology.uoc.gr/to-tmima/didaskontes/viky-foskoloy). She is also tutor in the module 'Arts I: Hellenic Fine Arts, Review of Greek Architecture, Urban Planning, and Urban Design' of the Hellenic Open University. Her research interests concentrate on the study of Byzantine monumental painting and iconography, on the interrelations in art between Byzantium and the West, and on aspects of private devotion in the Byzantine world: artefacts, beliefs, and practices with particular emphasis on the study of the cultural phenomenon of pilgrimage (see http://crete.academia.edu/VickyFoskolou).

Manolis Korres is Professor Emeritus of the National Technical University, Athens, member of the Academy of Athens, Seat of History of Ancient Architecture-Restoration, and president of the Committee for the Restoration of the Acropolis Monuments and of the Central Archaeological Council of Greece (Dr. Architect-Eng, Dr. Ph. h.c. FU Berlin). His work and research includes the Acropolis Restoration Project 1975, 1977–1980, 1982–1999, and architectural-archaeological research on Amorgos (1974–77), Naxos (1976–2000), Metropolis, Thessaly (2002–2005), Amyclae, Sparta (2006–), and Pythion, Thrace (1974–2008). Other subjects of study include Ionic capitals in Greece; Stoa of Eumenes, Monument of Philopappos, bridges of the Roman aqueduct, Temple of Olympios Zeus in Athens and Temple of Athena Pallenis, vernacular architecture in Mani, Etruscan Tombs in Norchia, the Theoderich Monument in Ravenna and many other. Scholarships include Gr. Ministry of Education, DAAD, Deutsches Archäologisches Institut, Princeton University, and Fulbright (1995). He has received the Bronze Medal, Academy of Athens (1989); Silver Medal, Académie d' Architecture, Paris (1995); Taxiarches of Phoenix of the Greek State, Premium Alexander von Humboldt (2003); and Gold Medal of the National Academy in Rome (Premio Internationale Feltrinelli, 2013). His publications include 12 books and 120 articles.

Lila Leontidou is an architect, planner, and geographer (Dipl.Arch. NTU Athens, M.Sc. LSE, PhD Univ. London), has been elected at different times as a member of the permanent academic staff in four Greek Universities and a British one (King's College London), the last post being at the Hellenic Open University (HOU – EAΠ), at which she is Professor Emerita. She has been twice the Dean of the School of Humanities and a professor of Geography and European Culture from 2002 until she retired. She was awarded the prize of academic excellence in 2017. She has been a Senior Fellow at the Hellenic Observatory of the European Institute, London School of Economics and Political Science (LSE), since 2012, and at the Johns Hopkins University since 1986, as well as twice a visiting professor in France (Université Paris 1-Panthéon-Sorbonne, Université de Caen) and until recently at the University of the Peloponnese for the English-speaking MA Mediterranean Studies (2014–8). An academic leader in several international research projects, she has published in Greek,

English, and French. Her work has been extensively cited and translated into Spanish, Italian, German, and Japanese. Her academic books include *The Mediterranean City in Transition* (Cambridge U.P. 1990/2006); *Cities of Silence* (in Greek, 1989/2013); *Geographically Illiterate Land* (in Greek, 2005/2011); the co-edited volumes *Mediterranean Tourism* (Routledge 2001) and *Urban Sprawl in Europe* (Blackwell 2007); and several other books which she (co) authored. The collective volume *Geographies in a Liquid Epoch*, authored by 34 academics (Propobos 2019), was published in her honour.

Argyro Loukaki is a professor of Greek Art, Architecture, and Urban Planning at HOU. She has a D.Phil. Oxford (School of Geography with Institute of Archaeology and Ashmolean Museum), MA Sussex (Urban and Regional Studies), Master Panteion (Regional Studies), and is Dipl. Arch. NTU Athens. She has received a Greek State Scholarship Foundation, Oxford prize, and Princeton fellowship. She has repeatedly directed the HOU Program 'Greek Culture and Civilization.' Previously, she was an architect/planner at the Greek Archaeological Service (design of urban and archaeological landscapes, etc.) and other public services (design for Kastella Hill, spatial planning for Methana Peninsula, etc.). Her publications include *Mediterranean Cultural Geography and Aesthetics of Development* (Kardamitsa 2007/2009), *Living Ruins, Value Conflicts* (Ashgate 2008, Routledge 2016), *The Geographical Unconscious* (Ashgate 2014, Routledge 2016), *Art and Space in Crisis-Hit Greece* (Leimon 2018, co-edited with Dimitris Plantzos), *Islands of the Ancient Sea: Aesthetics, Landscape and Modernity* (in press, Bloomsbury), and *Art, Space and Literature: The Case of Greece* (in press, Leimon). In 2017, she initiated a sequence of international conferences on art and space, now hosted every two years at the Acropolis Museum, Athens.

Konstantinos Moraitis is a professor at the School of Architecture, National Technical University (NTUA), Athens, Hellenic Republic. He has been teaching there since 1983. He is an architect-engineer (NTUA), and posseses a DEA in Ethical and Political Philosophy – Seminar of Aesthetics (Université de Paris I, Panthéon-Sorbonne, France) and a M.Sc in Arabic and Islamic Studies (Panteion University – Department of Political Science and History, Athens, Greece). His doctoral thesis was *Landscape: Allocating Place Through Civilisation. Exposition, and Theoretical Correlation of the Most Significant Modern Approaches Concerning Landscape* (NTUA). He is responsible for a postgraduate seminar on 'History and Theory of Landscape Design.' His publications include architectural projects, scientific articles, and authored tutorial books. He co-edited the volume *Urban Ethics Under Conditions of Crisis* with St. Rassia (World Scientific, 2019). He holds numerous distinctions in international and panhellenic architectural competitions.

Dionysis Mourelatos studied history and archeology at the University of Athens, where he continued his postgraduate and doctoral studies. He received his PhD from the University of Athens (2009). He has teaching experience on the graduate and undergraduate levels since 2009, as well as professional experience as an archaeologist at the Ministry of Culture, Greece, and in the Sinai Monastery. He has edited two books and is the author of various articles and chapters in edited volumes. He has participated in a large number of conferences with original papers.

Maria-Mirka Palioura studied French Letters and art history (BA, PhD Athens University, MA Université Paris I Panthèon-Sorbonne). She has edited two books and presented several conference papers on 19th-century Greek art. She has taught at the Athens School of Fine Arts, the Hellenic Open University, and worked in the Greek Ministry of Culture. She is a member of the Hellenic Association of Art Historians and is currently working in the Finopoulos Collection – Benaki Museum, Athens, Greece.

Dimitris Plantzos is a classical archaeologist, educated at Athens and Oxford. He is the author of various papers and books on Greek art and archaeology, archaeological theory, and classical reception. His Greek-language textbook on *Greek Art and Archaeology*, first published in 2011 by Kapon Editions, was published in 2016 in English and is now available through Lockwood Press in Atlanta, Georgia. His more recent books are *Archaeologies of the Classical* (Eikostos, 2014), an overview of archaeological method in the post-positivist era; *The Recent Future* (Nefeli, 2016), a study of archaeological biopolitics in contemporary Greece; and his study *The Art of Painting in Ancient Greece* (Kapon and Lockwood, 2018). He teaches classical archaeology and reception at the National and Kapodistrian University of Athens.

Konstantinos I. Soueref has been the director of the Ephorate of Antiquities of Ioannina at the Ministry of Culture and Sports since 2011 and was the director of the Ephorate of Antiquities of Florina and Kastoria (2006–2010). He holds a PhD from the School of History and Archaeology at the Aristotle University of Thessaloniki in Greece. He studied history and archaeology at the University of Bari, and he completed two postgraduate courses, in ancient philology at Scuola Normale Superiore of Pisa and in prehistoric archaeology at the University of Pisa, in Italy. He was appointed as adjunct lecturer and taught prehistoric archaeology at the School of Philosophy at the University of Ioannina (1989–1991), ancient Greek art at the University of Western Macedonia (2007–2011), and Greek arts at the School of Humanities of the Hellenic Open University (1999–2015). Dr. Soueref's scientific research focuses on the Mycenaean Civilization, the Greek Colonization, the Northern Greek Archaeology, and History of Art.

Erik Swyngedouw is professor of Geography at Manchester University. His research interests include urban-political ecology, hydro-social conflict, urban governance, democracy and political power, and the politics of globalisation. His was previously professor of geography at Oxford University and held the Vincent Wright Visiting Professorship at Sciences Po, Paris (2014), and a visiting Professorship at Wageningen University (2020). He recently co-edited (with Nik Heynen and Maria Kaika) *In the Nature of the City* (Rutledge), *The Post-Political and its Discontents: Spectres of Radical Politics Today* (with Japhy Willson) (Edinburg University Press, 2014), *Desalting the Seas* (with J. Williams), and *Water, Technology and the Nation State* (with F. Menga). He is the author of *Social Power and the Urbanization of Nature* (OUP 2004), *Liquid Power* (MIT 2015), and *Promises of the Political* (MIT 2018). He co-edited (with Henrik Ernstson) *Urban Political Ecology in the Anthropo-Obscene* (Routledge 2019). He is an elected member of the Academia Europaea and holds honorary Doctorates from Roskilde University (Denmark) and the University of Malmö (Sweden).

Anina Valkana studied history and archaeology at the National Kapodistrian University of Athens. She has a DEA in Art History from the Paris I-Sorbonne University. Her PhD thesis at the NTU Athens (Department of Architecture) is about the painting of Nikos Hadjikyriakos-Ghika, one of the most important Greek artists of the 1930s. She has been a curator at the National Museum of Contemporary Art in Athens since 2000. Her research interests focus on modern and contemporary art, art archives, documentation, and museum education. She has designed and implemented educational programs for different groups of audience. She has published on art history and museum education. She teaches art history and museum education at Harokopio University of Athens and art history at the Hellenic Open University from 2007.

Figures

Introduction

Part 1

Part 2

Part 3

Part 4

Maps and diagrams

Maps

***

Acknowledgments

The chapters of the book include, in revised form, a number of papers presented at the International Conference *Art and the City* which took place in the Acropolis Museum on March 17 and 18, 2017. The book proposal drew the immediate interest of the Routledge Editor Faye Leerink, who has been extremely prompt and kind throughout. The same is certainly true of Editorial Assistant Nonita Saha.

I proposed this conference as Coordinator and Professor of the Module Art-Architecture-Urban Planning of the Program 'Greek Civilization and Culture,' School of Humanities, Hellenic Open University, to the other members of the Organizing Committee. These included Tutors Dimitris Plantzos, Jenny Albani, Konstantinos I. Soueref, and Dionysis Mourelatos. The teamwork, which evolved in many lengthy meetings and myriads of frantic emails, has been sublimely wholehearted and trustworthy. Dr. Soueref helped us access the Acropolis Museum, where the conference was held. Dr. Albani and I shared many hurried and anxious moments at the end of the conference preparations. Throughout, she stayed a source of relaxing support. She also pursued additional sources of funding and of the graphic design of the conference materials. Dr. Mourelatos assisted with numerous practicalities. Professor Plantzos was there always. The scientific committee included the aforementioned, plus Professors Lila Leontidou, Konstantinos Moraitis, and Savvas Patsalidis. I am also grateful to the teaching group of the academic year 2016–2017, 25 people besides myself, for all their wonderful support.

Many thanks are owed to the Hellenic Open University (HOU) for the possibility to materialise the project, especially the President, the Vice President of Academic Affairs, and the School of Humanities which has sustained my accomplished efforts to turn a one-off event into an academic institution. Beyond the economic support, there have been all kinds of administrative and promotional support as well. I took a sabbatical between October 2019 and March 2020 for the completion of this and other authorial obligations.

I am absolutely grateful to Professor Dimitris Pantermalis, President of the Acropolis Museum, for hosting the conference in the museum's glamorous amphitheatre in such a generous, kind, and noble manner, a fact which contributed greatly to the effort. I also extend thanks to the Governing Council of the

Acropolis Museum and the personnel of the museum, especially Ms. Danae Zaousi, for all the collaboration.

Professors David Harvey and Erik Swyngedouw, main speakers of the conference, and, more importantly, precious friends since the Oxford years, supported this effort with their excellent speeches, their presence throughout the conference, and their fantastic participation in the festivities held in their honour.

Many thanks are due to all speakers of the conference and to a wonderful audience, which included, besides the general Athenian public, our students and colleagues from many universities. Interaction with the public was taken into account by all of us during paper revisions for this book.

Our secretariat during the conference, HOU's own Lila Kaldani and Foreini Zafeiratou, had a consistently graceful, angelic aura and the zest of those truly caring for a shared purpose. We were unfailingly assisted with warmth and goodwill by Afroditi Mesimeri, Head of the Athens Branch of the Hellenic Open University. Many thanks are also due to the enthusiastic HOU cameraman Panagiotis Fragoulis.

Dr. Andromachi Katselaki, Head of the Department of Educational Programs and Communication, Directorate of Museums, Hellenic Ministry of Culture, is thanked for undertaking the artistic design of the conference invitation, poster, and program, materialized by the graphic designer Spilios Pistas.

The Region and Governor of Attica and the Attica Distribution of Gas financially supported the conference.

My appreciation goes to the three anonymous reviewers for their generous and useful comments, which were taken into consideration in the Introduction in one way or another.

The authors had to respond to a demanding schedule and feedback process. The Academician and Professor Manolis Korres acted gracefully on our invitation to contribute with the foundational first chapter.

I want to thank my dear friends and colleagues at the School of Humanities of HOU, Vasso Hatzinikita and Konstantinos Petrogiannis, for their moral support during the implementation of this project and for those wonderful long discussions on everything under the sun our way to the city of Patras, seat of HOU, and back.

Last but not least, my gratitude and love are destined to my family, George and Lydia-Georgia, my mother, and my siblings, George and Angelos, for all the incredible support and the time 'stolen' from our preciously few moments together, spent on research and writing.

Introduction

Urban art between archetypal sublimity and ultramodern insurgency

Argyro Loukaki

Urban art and the city: creating, destroying, and reclaiming the sublime addresses the diachronic and vital relation between urban art and the city as a human-made sublime in the Eastern Mediterranean from classical times to current metropolises and their critical present. The city is seen as a work of art and the utmost collective creation of human energy, civilization, and culture. The past as a rich and unavoidable spatial unconscious emerges constantly, offering promise and hope despite the present crisis dynamics in the region. Current processes of contestation and redemption are linked to forms of the new 'imperialism of debt' and the positivist, technocratic understandings and demands of Euro-governments and neoliberal institutions, while evoking older forms of spatial power.

The urban space, whether real, symbolic, or artistically represented,[1] is here the object of apparent quasi-cycles and relevant processes of constitution, exaltation, imitation, contestation, and redemption through art, seen from a centuries-long historical perspective. New keys are proposed through the diversity and wide range of the chapters for the theorization and conception of the prismatic character of artistic reality as attributes of art, architecture and planning which correspond to the spectrum from urban glory to urban decline and back. This back-and-forth juncture is presently partly due to the globalizing, neoliberal condition; its detrimental effects upon countries and cities stir constant urban insurgences. These cause sequences of contestation as well as of inventive, artistic *dérives* by Guy Debord's standards. The humble means and objects, striking yet simple messages, particularly graffiti, theatrical happenings, and even the creation of impromptu urban gardens also shape an urban-scale *arte povera*.[2]

The growing interest in the subject area of this book, city and art, accrues from the important potential of beauty and culture as conduits of progress at times of widespread uncertainty and crisis. Especially given the present global and European situation, the book explores the theme of cities remaining creative and inventive under duress through urban art or art representing or recapitulating the urban space, offering theoretically informed and rigorously researched material on how politics, art and the city converge now.

Therefore, focal questions in this book are who defines, who implements, and who destroys urban beauty and sublimity? Who articulates moral judgments on the urban becoming in each and every temporal context? What is considered urban art in junctures like the present? Granted, art creates, disrupts, and recreates order by fighting apathy and by revealing the anguish of individuals and artists (see Scharfstein 2009, 435). But, should there be limits to urban devastation in the name of political struggle, beyond which dire damage is permanently inflicted to the collective psyche? Ultimately, to whom belongs the urban space now?

Awareness of the spatial is appropriated by art and artists, in acknowledged or unacknowledged manner, from notions and developments hatched in space-sensitive disciplines like geography, anthropology, architecture and planning. Therefore, this book specifically purports to establish a common ground between numerous relevant disciplines, including geography, history and theory of art and architecture, film studies, archaeology, monuments' restoration, landscape theory and design, cultural politics, criticism, and management, plus urban planning; authors' identities here often overlap. Major themes are the links that art, old and emergent, shapes between antiquity, politics, and the current crises in the region, plus the conscious or unconscious ways in which the past remains relevant as source of creative inspiration but also of urban and national identity.

The chapters of the book include, in revised form, a number of papers, presented at the International Conference *Art and the City*, which took place in the Acropolis Museum in March 2017; two additional authors were invited to contribute chapters (1 and 13). The intention has been to establish a realm for the study of the bonds between art and space in Greece, as well as globally.

The geographics of the book and the ancient aesthetic

The focal questions here indicate that a great number of major issues are posed as many limits and uncharted territories are tested – aesthetic, artistic, political, and activist – and quite a significant number of them are explored through theoretical and empirical materials. The approach, scale, and scope are global but also regional. However, this book cannot aspire to be an all-inclusive historical or geographical account of art in, and of, the city. In terms of geographical reference, the main focus is the Eastern Mediterranean. The time/culture spectrum covered is the sequence archaic, classical, Byzantine, and neoclassical urban glory, plus 20th-century modern urban ambitions and the current urban decline, the joint outcome of sustained early 21st century economic crisis and of public health hazards like the 2019/2020 coronavirus which highlight the importance of public urban space as necessary comforting and connecting factor. This is a circuit of constitution, attack, and reclaim of the sublime. A number of cities, impactful theoretically or empirically, figure in this volume, yet two cities are predominant: Greek capital city Athens throughout historical time, the most ancient, venerated seat of 'original' classical archetypes; and Byzantine Constantinople, a major cosmopolis, capital of Byzantium and of Byzantine art. Both have either been, or still are, part of the Greek geographical, imaginary, and symbolic space. Besides, a number of cities figure in this volume: Jerusalem, Ioannina, Thessaloniki, Candia (present-day Herakleion) and more. Space, whether more generic (the present insurgent city), mythical (Cythera), or imaginary (Andrei Tarkovsky's Zone), is juxtaposed with, but also introduces, theory, experience, and narrative, occasionally cinematic.

The theme of roots, belonging, and tradition is crucial to auteurs like Tarkovsky (1989, 93). The past as a rich, unavoidable spatial unconscious emerges constantly in this book, offering promise and hope despite the crisis dynamics. Recent developments have stirred the interest of the international press, including newspapers like the *Guardian* and the *New York Times*, which promote Athens as a dynamic pole of contemporary culture and as a looked-for international tourist destination. One of the reasons is the new official and spontaneous channels of artistic becoming which appear fresh, promising and flourishing despite the dire straits the country has been navigating for very long. On the other hand, while present processes of contestation and redemption are linked to the 'imperialism of debt,' they critically evoke older forms of spatial power.

Why the Eastern Mediterranean? Granted, it is one of many global fields of art and culture, ranging from Angkor to Teotihuacan, to Timbuktu, to Egypt's Giza, the Great Wall of China and so many more. Greek modern artists like painter Nikos Hadjikyriakos-Gkikas are aware of this, as seen in his letters around 1952 to his friend the eminent architect Dimitris Pikionis (Paisios ed. 1996), accounting that Chinese art includes exceptionally refined artefacts of a creative capacity hard to exceed, and that Mexican art is extremely authentic yet bloodthirsty and cruel (more on this later). Both creators were inspired by the purity of traditional Japanese art and architecture (Loukaki 2014/2016). So what exactly makes the uninterrupted and vast sequence of Greek cultures, at least since the Mycenaean Age of Heroes, echoing in Homer's epics, both unique and worth focusing on here? In other words, why the continuing importance of Greek classicism and the other Greek heritages (Figures I.1–4 and sporadically throughout the book), given the equalizing theoretical appraisal of the global cultural heritage voiced in international bodies like UNESCO? Add here the repeated efforts to shake classicism off while, ironically, imitating it – see, for instance, Picasso's dialogue with Greek classicism (Stambolidis and Berggruen 2019) – which reach what Boardman (1996, 13) names 'a fashionable bias against the classical' and Scharfstein (2009, 294) 'symbolic deaths and erasures of the past.' On the other hand, as the latter (ibid., 128) also notes:

> While I am not sure how the Apollo-experience influences our reactions to the art of other traditions, I think it has been bred into our bones for so long that it has become part of our visual nature. This remains true in spite of the passionate Western artistic revolt against tradition, because the passion of the revolt is still a kind of dependence.

The other issue posed here, accruing from the first, is how classicism is treated in the country that gave birth to it during the present difficult situation.

Quite simply, the fascination of classicism lies in the strongly anthropocentric focus which resulted in the exaltation of reason, reverent defiance of both the divine and human myth, as well as vigilant interest in aesthetic pleasure and

Figure I.1 The Athenian Acropolis from the Roman Agora. On the first level, the Tower of the Winds.

© Photo and source: Argyro Loukaki.

public participation, leading to the founding of the polis and of democracy. Perhaps this is what Schiller had in mind when speaking of 'the full humanity of the Greeks' (in Scharfstein 2009, 214). The union of myth, reason and human dignity continued to register in Greek art, theatre, architecture, literature, and philosophy, including systematic approaches to space, since at least the 8th century BC (Loukaki 2014/2016, 2019). Pre-Socratic thought explored the cosmic infinite as a higher, orderly mythological and ethical system to be imitated by mortals and their poleis, the utmost sacredness of light, mathematics, analogy, as well as the continuity between the human body, nature, and all manner of spatialities. Coupled with cultural spatialities from the entire Greek time-space spectrum, these dimensions have formed the basis of what I have proposed as 'Mediterranean cultural geography,' a geographical and ontological option of emancipatory and aesthetic spatial experience.

The visual has been axiomatically prevalent in this universe: 'The beautiful is beautiful as long as it is visible,' wrote poet Anacreon (582–485 BC). Passion for beauty and trepidation at the sublime are coeval with the birth of Greek poetry, epic and lyric, as well as of philosophical logos between the 8th and 6th centuries BC. Homer, fountainhead of poetry and myth narrative, but also, I would argue, of philosophical thought,[3] praised ecstatically the beauty of cities and humans alike. The sublime is likewise constantly present in Homer, causing delightful dread as divine epiphanies, frightful and majestic heroic actions, or cataclysmic events – descent to Hades, wild seas, terrible beasts, and huge fires. Through the sublime, Homer portrayed the Greek desire to transcend mortal limits toward assimilation of the human and the divine and to stamp collective memory and Greek spatial presence. The Homeric polis is sublime as the sacred height of human achievement, immortally and mortally constructed, divinely and humanly defended (see Scully 1990, 25–26, 93). Classical urban space (Chapter 1) was communally created, mythically represented, and poetically conceived. Poet Pindar praised Akragas (present-day Agrigento), the colonial city in Sicily established around 582 BC, as the most beautiful of the Greek world, 'inhabiting the hill of well-built dwellings.'[4] For all its sanctity, however, the polis was not inviolable. This classical paradox bespeaks the dignified, conscious human adventure resembling fleeting, sparkling firework arcs before the certain mortal darkness.

Nuances of urban beauty and sublimity allowed Greeks to propel the civilizing process within the compounds of the polis, both divine and secular: the heroic Homeric polis, home to relentless human passions, energies, and power was contained by the divine scheme and sperms of the democratic regime alike and stood in partial opposition to Sappho's intimate, quasi-domesticated, amorous sublimity (see Loukaki 2014/2016). Philosophy and urban planning were intertwined: the strongly political and spatial proclamation of philosopher Melissos (470–430 BC) that reality is homogeneous, eternal, infinite, and unitary is reflected in the urbanism of Hippodamos (498–408 BC) – a philosopher but also a physician, mathematician, meteorologist and architect – which aspired to a rational and just social order (Loukaki 2019).

Kant's aesthetics, classicism, and the present discontent

It accrues that art and the aesthetic are far from a brief brush with the desirous in human nature – though Kant's 'disinterestedness' does incorporate an idealist negation of the self[5] – or from purely personal taste. According to Kant,[6] who arrived at a theory of the judgment of taste wishing to explore the region beyond logic, the aesthetic not simply judges moral ideas using reflective analogy, but also cultivates a universal feeling of sympathy and the manifestation of the innermost self (Kant 1952, 226–227); this common threshold is the *sensus communis*. Why Kant here? Because he is the father figure of modern aesthetics; as such he is acknowledged, though not necessarily referenced by art historians (see later in this chapter). Moreover, he represents modernity's exploration of the aesthetic experience, including beauty and sublimity (see Elkins ed. 2006), first explored by the Greeks.

Although Kant drew from Burke and Longinus, he went substantially beyond them to encapsulate in the sublime and the beautiful the overall workings of the mind and the free play of the imagination.[7] Eighteenth-century aesthetics explored the multiple possibilities to imitate and express inarticulate human aspects through art (Guyer 2005, 16). This new, sharpened imagination was necessary, it seems to me, to grasp the possibilities open to European states through imperialism and colonialism. The kind of transcendent experiences and of new cognitive processes falling under 'the sublime' has been vital to the modernist adventure of the Western subject locally too, judging from the theory of trauma in Freud (on the former, see Wertheim 2006, 175–177).

The aesthetic discourse, defined as approach to the body, as stimulant of the imagination or as Kantian theory (Eagleton 1990), flows between philosophy and art history like sand in an hourglass. The sublime as infinity and tremendous natural might encompasses manifestations of the divine, the supernatural and enthralling in Kant, as in Homer. Encounter with the infinite and almighty does not leave human subjects impervious and unscathed; it may cause trauma, shock and stress, a sense of imbalance and self-loss, irresistible attraction and repulsion, awe and terror, which reach extremes such as violent fantasies of catharsis and oedipal anger against the 'parental' source (see Redfield 2006, 281–285). This shock affects both the spectator and the creator, conduit of the sublime, shows Walter Benjamin in his analysis of Charles Baudelaire's poetry. It also raises the soul above the 'vulgar commonplace,' as Kant named it, and propels the human existence beyond 'natural' limits.

Human labor can also reach sublimity if it overwhelms imagination and understanding (Wertheim 2006, 177). We already saw this in the Homeric polis. This condition intensifies when urban archetypes are restored and revitalized (Loukaki 2014/2016).

Figure I.2 The Parthenon in 1978. © Wikipedia Commons.

Photo: Steve Swayne. Source: Elaborated from https://en.wikipedia.org/wiki/Parthenon#/media/File:
The_Parthenon_in_Athens.jpg

Such is the case of the Parthenon, among the most quintessential, recognizable, and perfect architectural forms over global time and space (see Chapters 1 and 2).

Aesthetic terms, especially the beautiful and the sublime, aesthetic enjoyment or unease (Rancière's 'discontent'), and the aesthetics of defiance are strongly present, directly or indirectly, and variously elaborated in this book, beginning with its title. The new cultural life dictates new notions of the real for which a new critical and aesthetic language is needed and developing. Digitally created art propels a new visual sublime, an infinite, infinitely malleable, extremely dynamic and interactive system of visual information, new narratives, perceptions and visual expectations (Loukaki 2014/2016), which, however, is truly effective when landing, as it does now in Athens and other cities, in real urban experience. Consequently, both 'official' art and activist art are addressed. Part 3, in particular, variously explores the links among activist art, urban politics, the human body, classicism, and neoclassicism. Further, Chapter 10 pursues architectural and landscape theories and practices, alongside aesthetics and philosophy.

The triangle of aesthetics–politics–art is a field of active magma. To begin with, there is skepticism around aesthetic theory on the part of art historians and philosophers alike. Art historians and creators tend to circumscribe aesthetic judgments (see Elkins ed. 2006), preferring to abort Kant's categories from their focus. For instance, poet Giorgos Seferis (in Seferis and Tsatsos 1988, 82), a Greek Nobel laureate, was critical of philosophy's effort to control art using *a priori* classifications. Seferis argued that poets, despite their kinship with ecstatic mystics, must, unlike the latter, transmit poetic emotion to their audience. Artistic means for this transmission cannot be defined externally or in advance, because they shift constantly. But there has also been an 'internal' colonization of the aesthetic, as 19th-century philosophers like Schopenhauer and Hegel subordinated it to their idealistic philosophical systems (see Sejten 2016).

Further, a new boundary is now (re)rising, dwindling, arguably, the prospects of a global art history (see Elkins ed. 2007) posed by the present situation. Art still reflects what Braudel (in Braudel 1994, Chapter 3), based on Foucault, articulated as confines, conscious and unconscious, between civilizations. A global art history would demand the elaboration of aesthetic judgments and valuations of local contributions to the global cultural fund based on a set of global interpretive strategies which, however, is still missing, despite the role and presence of international conventions and cultural bodies such as UNESCO. Coming from the depths of an intuitive talent like Ghikas,' aesthetic judgments such as his (as shown previously) would today cause uneasiness. Current evaluative good manners tend either to manicure aesthetic argumentation, to safely limit it to Western works, or to approach immense independent national or regional traditions such as the Chinese through Western interpretive strategies because allegedly the West remains the almost sole province of modernity.

Western art history itself has bifurcated into a traditional emphasis on artistic autonomy, and into a socially sensitive analysis pioneered by other disciplines (Loukaki 2014/2016, 8). On the other hand, particularly since art started asking philosophical questions (Schürmann 2006, 183), art historians are contesting inherited disciplinary boundaries between art and philosophy, seeing their inter-action as an invitation to critical exchange (Rampley 2006).

'The art of non-art,' launched by modernism's visible aphorisms like Duchamp's (see Scharfstein 2009, 374), is also a field of potentially volcanic eruptions. Con-ceptual art like Duchamp's, particularly the provocative 'urinal' (see Chapter 7) contravenes two essential characteristics of 'the art of art:' First, its highest desire, which is nothing less than communion with the divine and expression of ultimate spiritual truths, according to Hegel. Second, the potential to be a masterpiece. Masterpieces are works epitomizing collective memories and emotions, linking past accomplishments with present innovation (see Kenneth Clark in Dell 2010, 8). The empowering role of public masterworks should never be underestimated: according to Tarkovsky (see Chapter 14), a sublime, purging trauma unites mas-terpiece and audience, uncovering the unfathomable depths of our own potential and the furthest reaches of our emotions (Tarkovsky 1989, 43). On the other hand, Duchamp's provocative anaesthesia toward Kant's judgment of taste, his ironic brand of fetishized mechanization has caused such ripples in art and politics as to help forge new bonds between them.

What is more, the various modern innovations revolutionized the way we see in general and the way we view and evaluate Byzantine art in particular (see Part 2 of this volume). After centuries of ambivalence, Byzantine art, a visual, mytho-logical, aesthetic and symbolic universe of high sophistication and vulnerability, suddenly emerged as cubism and surrealism *avant la lettre*, as I have shown else-where.[8] Byzantium, a reminder that quantification ultimately denudes the human experience, continues propelling amazing visual and spiritual refinements through high abstraction and sensuousness while still carrying the classical trace.

How did radical thought appropriate this opening that modernism offered it? With some notable exceptions from Walter Benjamin to Terry Eagleton via Ray-mond Williams, it felt uneasy toward the aesthetic-culture duet, which, however, has allowed European thought to articulate crucial social issues such as politics, ideology, the modern subject and scenarios for the future (Loukaki 2014/2016). Take the emphasis the 19th-century European bourgeois class placed on aesthetic education as a prerequisite for political empowerment. Plainly, aesthetic culture has been a political discourse assisting modernity's pursuit for social harmony, the consensual ground for representative democracy (see Redfield 2006).

The French radical thought has been more eager to contribute toward a rap-prochement between philosophy, art, and politics. To Alain Badiou philosophy binds through logic the operations of politics and art; Rancière (in Rancière 2009, 66) concurs, identifying art with non-art. The pair of politics–art is archetypal, as we just saw. However, urban art in classical Greece pursued the glory of the city (see Chapter 1); it was not damaged while the country was being brought to its knees, as is partly witnessed in Athens today (see Chapter 2).

Alain Badiou's (2004, 4) claim that to Aristotle art pertains to the ethical, not the theoretical, contravenes the precepts of Greek philosophy and urban planning previously discussed. Badiou also distinguishes three different manners in which philosophy is tied to art[9] somewhat unidimensionally. For instance, his view of Plato as exclusively didactic irons out the exquisite contradictions of a great philosopher but also poet like Plato, passionate about the role of poetic mania, as seen in *Phaedrus* (verses 249–252):

ὃς δ' ἂν ἄνευ μανίας Μουσῶν ἐπὶ
ποιητικὰς θύρας ἀφίκηται, πεισθεὶς ὡς ἄρα ἐκ τέχνης ἱκανὸς
ποιητὴς ἐσόμενος, ἀτελὴς αὐτός τε καὶ ἡ ποίησις ὑπὸ τῆς
τῶν μαινομένων ἡ τοῦ σωφρονοῦντος ἠφανίσθη.

But he who without divine madness
comes to the doors of the Muses, confident that he will be a good poet by art,
meets with no success, and the poetry of the same man vanishes into
  nothingness
before that of inspired madmen.

Badiou likewise understates how Platonic ideas helped liberate art (see Hofstadter and Kuhns 1976, xv) as well as the Platonic identification of wisdom with utmost beauty (in *Symposium*). The same aphoristic treatment is reserved for Aristotle who, according to Badiou (ibid.), focuses on the cathartic in tragedy, not at all on the cognitive or revelatory. But Aristotle also explores spectators' emotional participation through the sheer beauty and sublimity, the pleasure of tragedy[10] (*Poetics*, Chapter 14; see also Schaper 1968). Besides, his 'familiar pleasure,'[11] emerging from the cognitive value of tragedy, approximates the universal pleasure of learning and wisdom. In other words, catharsis, originating in mercy and fear, is not the only, but the *last* moment of a climaxing sublime experience.

To summarize, discussion here takes place against a number of givens: first, there is a highly precarious complementarity between public traditional and activist artforms. Second, classicism has been treated elliptically as public art and as a nuanced cultural project against a background of admiring yet relentless contestation (Figures I.3 and 2.3). Classicism is important here as the book deals with some new artforms like insurgent street art and digital art in parallel to, and contrasted with the archetypal classicism of Athens, but also with recent and current painting artworks focusing on this city. The third given is, more generally, that links between art, politics, aesthetics, and philosophy define a tensely fluid arena.

The challenge is always to turn this tension into creative expression for the many. This is imperative because both theoretical research and awareness of the recently imposed 'constant indebtedness' on certain countries and cities support notions of the urban space as the object of apparent quasi-cycles, as indicated in the title of this book. Hence, this exploration contravenes the belief of architects Herzog and de Meuron (2008) that there are no theories of cities, there are only cities, which ignores the extraordinary theoretical erudition accumulated on cities, their geography, sociology, culture, anthropology, architecture, urban planning, history, or economy.

Figure I.3 Classicism contested in its cradle. Tugging in the wall of the Ancient Agora, across the Acropolis, speedily removed in early 2020.

© Photo and source: Argyro Loukaki.

Figure I.4 Simultaneously, sublime urban experiences are constantly reclaimed next to the Acropolis and the garden of modern architect Dimitris Pikionis.

© Photo and source: Argyro Loukaki.

Greece's present ordeal and 'saving the European phenomena'

Current urban political struggles happening in the world at the moment are regularly associated with the impacts of neoliberalism and as ferocious as they are manifold. The condition of urban art in countries like Greece is intricate exactly because of this circumstance. But what is exactly the impact of constant indebtedness that Greece, a country with a deep sense of traumatic loss, controlled by its creditors for the next hundred years, has been experiencing?

This country is plunged into one of the deepest and longest depressions in economic history according to *Financial Times*,[12] and at a time of peace. Some data are necessary here to capture the extent of damage inflicted on Greece, at once economic, social, and in terms of national prestige. A 'rectifying' bailout – to many prestigious economists entirely unnecessary and deleterious[13] – was matched with, and facilitated by a concomitant imperative priority, defined with a deeply anti-democratic reflex and utter disregard for the impacts on the Greek people: to give the Eurozone time to build a firewall limiting contagion. Much of the money lent to Greece went to allow lenders to escape, on the excuse that this was a 'systemic' crisis. Yannis Varoufakis (2017), former Minister of Finance, has shown that Greece's democracy has been met with callous disregard; its economy and sovereignty with collusion, hypocrisy, a punishing spirit and constant blackmailing on the part of Europe's political, financial and media elite. The bailout terms were defined by the 'troika,' later 'institutions,' the European Commission, the European Central Bank (ECB), and the International Monetary Fund (IMF). The latter has disclosed 'notable failures' in the initial bailout programme including a deeper-than-expected recession, a run on the banking system, and exceptionally high unemployment which has continued unabated. Greece has been constantly repaying considerable amounts throughout this period to the ECB, the IMF and to private creditors.

The possibility of upfront debt restructuring was rejected by the European Commission; the programme did not restore growth as it had set out to do. Actually, not a single measure has targeted the restoration of the Greek economy. Efforts focused on saving the banks, while the Greek people were met with unyielding cruelty and intransigence. Suicide statistics increased by 33% between 2009 and 2015.[14] In 2016 there has been an unprecedented delegation of the movable and immovable property of the country until 2115, again without national democratic control or accountability through a 'Super-Fund' which also got access to archaeological sites like Knossos, archaeological museums like the Herakleion museum, ancient theatres, and fortresses.[15]

Greeks have already given to the belt mechanism of this Eternal Debt Slavery 52% of pensions, 38% of salaries, more than a million unemployed, almost the whole of public utilities, the value collapse of almost all private estate.[16] The Gross Domestic Product (GDP) decreased 28% from the beginning of the crisis, the Gross National Income (GNI) from 93% of European median in 2008 diminished to 67% in 2017. But despite more than 10 years of bloody austerity, the national debt has escalated from approximately 130% to approximately 190%. There is a swift expanse of extremely low salaries for the 'flexible' works on offer. Still, the country must unequivocally achieve a primary budgetary surplus of 3.5% and therefore must continue brutal policies of public spending reduction in investments and in all aspects of the welfare state. As a result, the health system deteriorated and the health standard decreased in the last decade. People could not be fed satisfactorily according to a study of 2016[17] and the number of homeless soared, as can be witnessed in the arcades of central Athenian boulevards like Stadiou Street (see Chapter 2).

These 'years of stone' have accumulated a lot of suffering. 'They (the Euro-group) saw Greece as a laboratory of extreme austerity and faced its people with the same apathy guinea pigs are treated,' according to movie director Kostas Gavras.[18] Chapter 2 discusses in passing the relevant phenomenon of victimizing and scapegoating. Greeks are strained with taxes and surcharges which increased faster than the GDP in 2018.[19] Unemployment statistics are still the worst in the European Union according to Eurostat.[20] This massive reduction in the standard of living is theoretically expected to rise to pro-crisis level in around 2035. If Greek society has not entirely collapsed, it is because of the social shield provided by strong family bonds.[21]

But there are more impacts. Greece faces the ticking bomb of the demographic imbalance and shrinkage. There is fast ageing of the Greek population, decrease of the population after 2010 because of the reduction of fertility due to the economic situation, and increase in outmigration, as around 600,000 young and highly educated people migrated to Western countries to make a decent living (the Greek 'brain drain'). In 2018 there were 38,000 more deaths than births. In 2035 the population will be 1.4 million people less that what it was in 2015, when the population rose to 10.9 million people.[22]

A further critical problem is the influx of economic immigrants from many countries alongside Syrian refugees. This issue of global mobility, which has challenged the people of Europe, obviously supersedes national policies and has increasingly raised urgent questions of national and local security. Apostles for opening up fortress Europe like Jürgen Habermas turned their attention to the worries of the German public opinion; immigration has been among the reasons for Brexit, uniting anti-globalists, Euroskeptics and nationalists. Analysts speak of a new globalized market despotism imposing wandering nomads on the basis of an ideologically validated free labour movement and of money flows involving NGOs. Greece, as the frontier of the European Union with Turkey, has faced a dramatic increase in arrivals of refugees and illegal immigrants reaching the anguished islands of the Eastern Aegean (which initially welcomed them whole-heartedly) in waves of hundreds and thousands daily, becoming an immense national hotspot.[23] After the 2015 migrant crisis,[24] when many sea entrants headed to Germany via Greece, the country has been left practically alone to deal with severe humanitarian and geostrategic impacts,[25] as migrants have been treated as pawns or as downright battering rams against the Greek borders, simultaneously the European Union borders. Things were culminated with Turkish provocative actions in early 2020 orchestrating massive, aggressive swarms to cross Greek land and island borders; these moves were officially deemed hybrid attacks targeting the destabilization of Greece.[26]

'Spaces of hope'[27] moulded with art as conduit of beauty and sublimity are essential to this book. But there is a price for their establishment. Crisis-hit Greece has been left virtually on its own, either as lateral shock absorber, or as 'saver of the phenomena,' the civilized humanity of Europe, at great national cost.

The thematic cycles

Given the aforementioned context, which, deeply affects current forms of urban art in ancient Mediterranean countries but does so always in deep conjunction with their pasts, the thematic cycles of the book are constituted in approximately linear chronological sequence. However, sequence is interrupted by themes recurrent or associated throughout time; classicism is a constant springboard, archetype, and human-made sublime – especially when archaeological sites are incorporated into the modern urban tissue through landscaping[28] – recurring in ways similar to the associative workings of human memory. The past is here a profuse, complex and multifaceted unconscious both geographically and temporally constantly reflected in, but also challenged by the present. Contrasting this richness, the book elaborates on the present condition and aesthetic of 'the least' which has a deep local poetic and philosophical origin as well as a compulsory character under the present circumstances. These recurrent themes are subject to elaboration from new viewpoints. Spatial attributes of art and architecture, artistic activism and Greece in crisis are themes already explored (see Foster 2011; Summers 2003; Tziovas ed. 2017; Loukaki and Plantzos eds. 2018; Luger and Ren eds. 2017). This volume differs on account of both its subject and time horizon. However, the aforementioned books have usefully paved the way. Throughout, theoretical accounts and empirical approaches balance out. This happens both among chapters as well as in the same chapter. Relevant bibliography on quite the subject matter of the book at hand is scarce, therefore it is hoped that it will contribute toward filling a sensed lacuna.

Urban architecture is art often and diachronically (Chapter 1), not just now, as is sometimes claimed. On its part, art is increasingly dependent on architecture as its adjunct – just consider public and private architecture as bearer of all kinds of art; this is especially apparent in a contested artform, street art which flourishes now for better *and* worse. Present processes of contestation and redemption, which include multifarious instances of 'creative destruction,' are closely linked to forms of constant debt regimes as direct or indirect responses to it.

The book includes four thematic cycles addressed in equivalent parts:

Part 1 *Creating, Imitating and Destroying the Urban Sublime and its Materiality: From Athenian Classicism to Neoclassicism and the Present* covers the first thematic cycle. This part presents innovative aspects of both the ancient Greek idea of urban beauty and sublimity, and the rhizomatic, conceptual importance of the materials cities are made of. Exploration expands (spatially) horizontally to architecture, sculpture, and urban planning, and (temporally) vertically to the founding, imitation, and current contestation of classicism. Athenian classicism was appropriated and imitated by successive cultures of what presently belongs to the Western sphere, local and foreign, including the Roman and Christian Athenian, the romantic/neoclassical, and the Modern Greek. The entwined expression of ethical and aesthetic ideals, it should be noted, challenges Badiou's propositions discussed previously. Athens was chosen as capital of the Greek state by King Ludwig of Bavaria (Bayern) in 1832 mainly due to its classicism, subsequently imitated by neoclassicism, mentioned here in Chapter 2, and thereafter in Chapters 8, 9, 12, and 13. The syntactic importance of the latter and the close imitation of classical prototypes both in terms of morphology and of material, mainly Pentelic marble, facilitated by immediate proximity, make Athenian neoclassicism quite unique.

Chapter 1 (Ancient Greek Cities as Works of Art), authored by Manolis Korres, explores in foundational manner the formation of ancient Greek cities as works of art and beauty in the archaic (from the 8th century BC; see Map 1), the classical, the Hellenistic and the Roman period, examining idealistic and utilitarian elements. Both citizens' lives and the arts were highly public; the latter served utmost artistic ends. Buildings' arrangement and artistically fashioned open spaces of various forms were linked to the city plan which bore witness to underlying concepts and artistic intentions. Combination of various parameters, including geometric arrangements and patterns, symmetries, numerical proportions, and repetition of shapes resulted in numerous particular cases. Korres argues that the artistic theatricality and the figural decoration of public architecture transformed

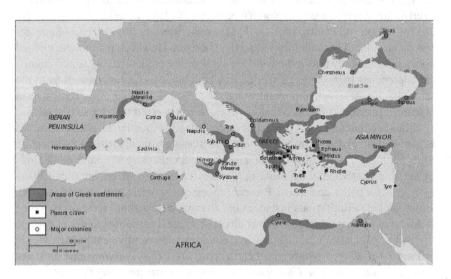

Map 1 Areas settled by Greeks by the close of the archaic period.

© Wikipedia Commons. Source: https://en.wikipedia.org/wiki/Archaic_Greece#/media/File:Greek_Colonization_Archaic_Period.svg by Dipa1965

many ancient cities into vast outdoor art exhibitions, of which the current archaeological museums, even the richest, are not but faint reflections.

Chapter 2 (Assaulting the Archetypes: Urban Materiality and the Current Adulation and Hatred for Marble in Athens), authored by Argyro Loukaki, explores a current Athenian paradox: marble architectural and sculptural monuments have been treated with an 'irrational' blend of collective adulation and activist rage. The complex effects of archetypal materiality are explored through ancient philosophy, art, neuroscience, architecture, psychoanalysis, cultural geography, archaeology, literature, and geology. Penteli, the Athenian marble mountain, which provided the material for the Acropolis monuments and was incorporated into classical culture as location of Plato's 'cave' and of the birth of drama by Thespis, now bridges the ancient and modern city. The material rootedness of Athens emerges thus as an engine of social coherence, collective meaning, and the urban imaginary; activist mutilation of the urban materiality disenfranchises the social body.

Part 2 *The Artistic Sublime of the Byzantine Cosmopolis: Between Imitation of Classicism and the Christian Dogma* introduces the second thematic cycle, which follows the current trend for emphasis on space perception and importance, applying it in Byzantine art and studies. There are coeval as well as modern readings and representations of the Byzantine urban space; both are relevant here. For artistic, philosophical, juridical, scientific, and architectural patterns, Constantinople constantly referred to Greek, especially Athenian, and Roman classicism; however, its religious reference was Jerusalem, seat of the Christian religion, thence of sacredness. Space was closely linked with the exploration of holiness

Figure I.5 Athens and the Acropolis overlooking the Saronic Gulf and the islands Salamis to the right and Aegina to the left of the background.

© Photo and source: Argyro Loukaki.

through polar opposites such as heavenly (Jerusalem, Constantinople) and sinful (Babylon). Constantinople's perception through literary and visual means was monitored in ways that would ensure aesthetic and material reproduction of a great city and its empire. Texts, images, processions, and visual conventions crafted carefully versions of ideal spatialities through imitation.

Part 2 also reveals some little-known processes of cultural exchange between East and West. The themes explored here may link Byzantine studies with modern urban studies, allowing for comparison on the means of sponsoring certain major urban ideas to the exclusion of others.

Chapter 3 (On Real and Imaginary Cities. Textual and Visual Representation of Cities and the Perception of Urban Space in the Byzantine World), authored by Vicky Foskolou, argues that Byzantine cities are explored in terms of their transformation, fortification, or the urban planning of Constantinople. Their textual and visual descriptions usually offer idealized images: surrounded by high walls, Byzantine cities loom in the middle of fertile natural environments, decorated with important buildings. However, textual and visual evidence rarely reveals pictures of the everyday, trivial, or even dark side of cities. The aim of Foskolou's study is to distinguish between sacred and profane urban spaces, looking behind the 'brilliant' representations of texts and images, and to contribute toward the further discovery of everyday urban life in Byzantium.

Chapter 4 (Art as Carrier of the Identity and Reflection of the Great Polis (Constantinople) and the Sacred Polis (Jerusalem) in Byzantine Provinces), authored by Charalampos G. Chotzakoglou, stresses the importance of (Megalo)-Polis (Constantinople) and Hagia Polis (Jerusalem) as spiritual, topographical, architectural and artistic models, which were imitated by Byzantine provinces and western towns, either as *loca sancta*, toponyms, urban planning, artistic representations, architectural formations, ritual ceremonies, or as philological *topoi*. The Constantinopolitan planning model and monasteries but also the church of Hagia Sophia (Holy Wisdom) were imitated in the provinces. The author uses various original examples, especially from Cyprus, to show how specific landmarks entered the Byzantine art as visual symbols and reflections of the Byzantine Oecumene and its two main urban models.

Chapter 5 (Depictions of the Virgin in an 11th-century Panel at Sinai as Perception of the City of Constantinople), authored by Dionysis Mourelatos, discusses how the use of Marian icons in public Constantinopolitan ceremonies influenced the perception of the city. It is argued that in 11th-century Constantinople, the church of Hagia Sophia and a number of Marian miraculous icons were the most representative relics. The chapter focuses on a hexaptych from Sinai monastery, which includes depictions of the four most venerated such icons. Hagia Sophia's apse was decorated with a figure of Virgin Mary with child Christ which was of the same type as on the Sinaitic panel, whose donor was a Sinai monk; this, according to Mourelatos, may epitomize his perception of Constantinople as a Marian city.

Chapter 6 (Between Convention and Reality: Visual Approaches to the City in Post-Byzantine Icon Painting), authored by Jenny Albani, begins from the finding that the Byzantine East is not familiar with realistic visual representations of the city. However, Albani claims that cityscapes in some post-Byzantine icons are both background motifs to religious scenes and visual expressions of the donors' notions of a specific urban social character. The rendering of cities conforms to visual conventions of Byzantine art discussed in the text. Elements of the visual narrative, like inscriptions, contribute to the visual articulation of criticism towards the city in 'Byzantium after Byzantium.' Based on a number of examples, the chapter aims to a descriptive and interpretative approach of 16th-century icon painting by means of art-historical comparanda and texts.

Part 3 *Current Contestations of the Urban Sublime: Crisis and Urban Insurgency through Art* covers the third thematic cycle, which explores art, aesthetic, and experience in the current urban space of Athens, which sets the tone for national and local evolutions. As the Greek capital city, Athens is in many ways an urban guinea pig, bearing the scars of constant conflicts. While the city is turned into an epicenter of revolt, the limits, strengths and carrying capacities of Athenian and national space are direly tested in ways mentioned already. Contested forms of expression emerge, which often move across borderline forms of neo-*arte povera* (graffiti, impromptu theatrical performances) and irate destruction. On the other hand, Athens is now also at the epicenter of global cultural interest.

This part juxtaposes a more comprehensive geographical discourse and insurgent urban artistic happenings in this city. Here are included outlooks of urban art linked and contrasted to dimensions of politics, the classical unconscious, visibility, tactility, ideology, plus transcendence, creative destruction, the human body in the city now, as well as artistic traces of the ephemeral.

Chapter 6 (Urban Insurgency as Political Art), authored by Erik Swyngedouw, argues, drawing from Alain Badiou, that the art–politics link resides in the articulation between these two social realities as distinct truth procedures and aesthetic categories, understood in Rancièrian mode as 'the distribution of the sensible.' Artistic and political interventions signal in their performative staging the inegalitarian forms of the present, argues Swyngedouw. The truth of art remains singular, while the truth of politics is universalizing. Their conceptualizations in this chapter reveal how urban art and emancipatory urban politics intertwine. It is claimed that the proliferation of urban rebellions disturbs the neoliberal status-quo plus various economic and political elites because it exposes the variegated 'wrongs' and spiralling inequalities of the current autocratic neo-liberalization.

Chapter 8 (Athens, Invisible City: From Neoclassical Re-construction to the Dystopia of the Crisis and its Contestation by the Urban Grassroots), authored by Lila Leontidou, explores multidimensionally the interaction between 'invisible cities' (Calvino) and visibly material ones; both reflect urban identities through art and architecture. Athens expresses quite this interplay (Raban). The chapter presents 19th-century neoclassical architecture; discusses the illusions of the 2004 Olympic Games; and concludes with the city of crisis and austerity, which the author analyses by means of the concepts of 'Orientalism' (Said) and 'crypto-colonialism' (Herzfeld) applied in the present-day European Union. The dystopic 'soft city' imposed by money lenders nourished the grassroots 'movements of the piazzas' of the period 2011–2013. New artistic expressions, created by solidarity networks in hybrid spaces of the digital society enhance the distinctive palimpsest of Athens.

Chapter 9 (Bodies in the City: Athenian Street Art and the Biopolitics of the 'Greek Crisis'), authored by Dimitris Plantzos, explores the neoliberal biopolitics and the new corporeality in Athens as an insurgent city which have emerged from the onset of the 'Greek crisis' in 2009, in which classicism is shown to be variously entangled. It is argued that classical antiquity, which has always affected Modern Greek identity, has been used in the context of the 'Greek crisis' both as an emancipatory tool against the country's debtors as well as a disciplinary device on behalf of its critics. The chapter discusses new artistic media, mainly Athenian graffiti from the neighbourhood of Exarcheia, incorporating themes from classical Greek art to express their makers' sentiments and the frustrations of their public at large.

Part 4 *(Re)-Constituting the Sublime City: Nature, Sculpture, Architecture, Film and Politics in Representations of Modern and Ultramodern Space* hosts the fourth thematic cycle which explores the creative process of re-constituting the urban sublime. This process propels a perspective of transcendence, placed under visual and poetic elaboration on an aesthetic, symbolic and psychoanalytic level. Part 3 focused on contesting creativity. Here, the dynamic development of the arts moves towards reconstituting creativity. Exploration of sublime landscapes is a backdrop reference to such urban processes that have already taken place or are taking place right now. The existential and mythical aspects of the present have been cinematically explored by Tarkovsky and Angelopoulos in ways accounted for here. Both creators, it should be noted, use cinema to explore the history and culture of their countries. Tarkovsky's book and universe transmit a deep spirituality, reared in the Russian Orthodox tradition; Angelopoulos carries the whole of Greek and Balkan tradition, including Byzantine iconography and ceremony, claims Horton (1999).

Part of this world (and, incidentally, setting for an Angelopoulos' film), the urban space of Ioannina in north-western Greece, a city with a rich cultural palimpsest, appropriated an important modernist architectonic and sculptural vocabulary to redefine its urban future and buttress its Greekness. Previously, early 20th-century Athens created horticultural and sculptural micro-landscapes to honour great heroes or politicians. Right now, it transpires that it is urban art and culture, not the economy and the positivist thinking of European bureaucrats that promises possibilities of recovery as optimistic symbolism, public awareness, and innovative development. Weakness is turned into an expressive dynamic through the drive of creative 'translators,' as it happens with both impromptu urban gardens and jungles, real and in graffito form, and with urban scenes of two accomplished painters.

Chapter 10 (Urban Gardening as a Collective Participatory Art: Landscape and Political Qualities Related to the Concept of the 'Sublime'), authored by Konstantinos Moraitis, explores the sublime, associated with natural magnificence and religious feeling, as an object of ideology, drawing, among other theoreticians, from Taut, Lacan, and Žižek. On a practical level, the chapter discusses a jungle scene graffito in Exarcheia, Athens, which announces the local demand for 'green urbanism'; further, it deliberates on the art of urban gardening, correlated to the 'political sustainability' of urban ethics and to the democratic principles of Western societies. Moraitis additionally focuses on the priority of environmental amelioration and on idealized landscape paradigms of promising Arcadias, with reference to emblematic cultural landscapes of the Hellenic and Roman antiquity, to indicate the political aspects of gardening as urban sublime.

Chapter 11 ('Trikoupis Refuses to Unveil Himself in Order not to See': A Memorial Statue and National Identity in Early 20th-Century Greece), authored by Maria-Mirka Palioura, discusses official politics of memory in the Athenian space by means of public sculpture. Palioura chronicles the efforts to set up a monument honoring Charilaos Trikoupis, an important politician, before the then seat of the Greek Parliament, dated in 1896. Also, the subsequent move of this monument to a less prominent place, which freed space for the equestrian statue of General Theodoros Kolokotronis, the hero of the 1821 Greek War of Independence. The double act of relocation reflects the symbolic narrative of communal and national identity at a time when the Greek Struggle for Independence was deemed far superior to Trikoupis' contribution.

Chapter 12 (Dialogues with Modernity in the City of Ioannina: Aris Konstantinidis, Natalia Mela, and Paris Prekas), authored by Konstantinos I. Soueref, addresses the following questions: Why did the city of Ioannina of the 1960s and 1970s choose to glamorize through the architecture or the sculpture of, among others, Aris Konstantinidis (the archaeological museum on Litharitsia Hill), Natalia Mela (the statue of 'the Hoplite'), and Paris Prekas (a relief composition for a bank façade)? Is this effort still relevant, and has there been a move from an established attachment to local material tradition toward other forms of art and technique since the 1960s? These artistic interventions in a regional city, placed beside the ancient, Byzantine and Ottoman local heritage of Epirus, constitute a cultural revolution, according to Soueref.

Chapter 13 (Painting Versions of the Athenian Landscape: Spyros Vassileiou and Yiannis Adamakis), authored by Anina Valkana, approaches 'romantic' versions of the Athenian landscape in the postwar period and during the present crisis through the works of painters Spyros Vassiliou and Yannis Adamakis. Though commencing from diverse ideological and aesthetic backgrounds, and belonging to different generations, the work of both is dominated by the Athenian urban landscape. Urban symbols activate the viewer sensorily. In Vassiliou, natural surroundings are merging with architecture, history, and memories of urban life. In Adamakis, atmospheric biographies of buildings imply the new status of Athenian residents, suggesting a poetic, psychoanalytic experience of the urban space.

Chapter 14 (The Mythical Landscape of Andrei Tarkovsky: Notes on the Interpretation of Cinematic Space in *Stalker*), authored by Stavros Alifragkis, is a close study of Tarkovsky's *Stalker* (USSR, 1979), which explores the construction of a distinctive filmic landscape, the post-apocalyptic Zone. The questions posed are:

How framing and editing affect the audiences' perception of physical space, and how different shooting locations merge into one, meaningful and seamless filmic space, becoming place. Tarkovsky's mise-en-cadre, mise-en-scène, and continuity editing provide the backdrop for this investigation; he is also compared with Greek director Theo Angelopoulos. References include Lynch's imageability, Norberg-Schulz's *genius loci*, Foucault's heterotopia, and Bachelard's 'dialectics of outside and inside.' These notions describe key spatial phenomena of supermodernity, as introduced by Augé.

The book closes with some concluding thoughts.

Notes

1 There is a lengthy bibliography on space ontology. See Lefebvre (1992); Harvey (2006); Loukaki (2014/2016).
2 *Arte Povera*, Italian for 'poor art,' was an influential avant-garde movement that emerged in the 1960s using cheap or worthless materials.
3 Loukaki (2014/2016, 2019).
4 See also Chapter 1.
5 Either the aesthetic experience motivates the imagination and conception, or this is about a sensual satisfaction (Scruton 1979). But when admiring divine greatness, calm reflection is required (Kant 1952, 103–113).
6 In *The Critique of Judgment (1790/*1952) Kant introduced or developed most of the critical terms in modernist art analysis, beauty and the sublime, reflective judgement, intellectual pleasure and free play of cognitive and imaginative elements in aesthetic appreciation and more, including the *sensus communis* (Gero 2006, 5).
7 It includes both the content and formal dimensions of representation: order, diction, style, materials, harmony, composition, instrumentation.
8 Loukaki (2007/2009, 2014/2016, 2020).
9 Didactically, in Plato's way, which views art's purpose extrinsic to truth (art is the imitation not of things but of the effect of truth), romantically (only art is capable of truth), and classically, in Aristotle's way (art is incapable of truth, its essence is mimetic; its role, 'catharsis,' is therapeutic, not cognitive or revelatory).
10 https://eclass.uowm.gr/modules/document/file.php/EETF189/ΑΡΙΣΤΟΤΕΛΗΣ.pdf (Accessed 20.10.2019).
11 *Poetics*, 1459, 521.
12 Wolf (2019).
13 See, for instance, https://slpress.gr/oikonomia/poso-kostise-i-troika-stin-ellada-me-noymera/
14 www.kathimerini.gr/1067900/article/epikairothta/ellada/ay3h8hkan-kata-33-oi-aytok tonies-sthn-ellada-sthn-periodo-ths-krishs (Accessed 06.03.2020).
15 www.kathimerini.gr/987328/article/epikairothta/ellada/h-ekpoihsh-mnhmeiwn-kai-oi-ey8ynes (Accessed 25.10.2019).
16 https://slpress.gr/politiki/to-politiko-systima-katapinei-tin-koinonia-kai-ton-eayto-toy/ (Accessed 19.01.2019).
17 Tountas ed. (2016).
18 www.lifo.gr/articles/cinema_articles/256244/misel-gavra-peninta-xronia-meta-to-z-ta-idia-sk-apo-ton-typo (Accessed 30.10.2019).
19 www.economistas.gr/oikonomia/21002_i-ellada-kanei-protathlitismo-se-foroys-kai-eisfores (Accessed 24.09.2019).
20 https://ec.europa.eu/eurostat/statistics-explained/index.php/Unemployment_statis tics_at_regional_level (Accessed 24.09.2019).

21 www.kathimerini.gr/1041633/article/epikairothta/ellada/h-ellhnikh-oikogeneia-ante3e-sth-megalh-krish (Accessed 25.09.2019).

22 www.kathimerini.gr/1013112/article/epikairothta/ellada/voylh-dhmografiko-narkh-gia-synta3iodotiko-kai-systhma-ygeias-h-ghransh-toy-plh8ysmoy (Accessed 28.07.2019).

23 Only in October 2019 there were 6.868 arrivals in these islands.

24 There were over one million sea entrants reaching a monthly peak of over 221.000 in October of that year, see www.unhcr.org/news/latest/2015/12/5683d0b56/million-sea-arrivals-reach-europe-2015.html (Accessed 06.03.2020). European solidarity has amounted to little beyond the establishment of Frontex, which has been of limited effectiveness, and the funding of NGOs – some of which have been involved in decision-making on geographical and shelter allocation ignoring local authorities.

25 According to official statistics, in late January 2020 there were 115,600 refugees in Greece, 74,400 inland and 41,200 on the islands. They have profited from the program HESTIA for housing, health security, education, and work; see https://data2.unhcr.org/en/documents/download/74391 (Accessed 07.03.2020).

26 www.europarl.europa.eu/doceo/document/P-9-2020-001224_EN.html and www.msn.com/en-gb/news/world/refugees-told-europe-is-closed-as-tensions-rise-at-greece-turkey-border/ar-BB10QtcW?li=BBoPWjQ#image=5 (Accessed 06.03.2020).

This humanitarian devastation is non-manageable according to Greek authorities, given the European denial to shoulder proportionally any load of non-monetary and non-material nature. Combined, actions of unsolicited NGOs, infrastructural demands, distressing impacts – economic (e.g., tourism profits), land-use, social, and cultural – the excessive length of hospitality, extreme inhabitant-immigrant ratios in sensitive border spaces, topped with geostrategic plans of aspiring powers in the area have tested every manner of corresponding, fragilised Greek carrying capacities.

27 See in various chapters here references to David Harvey's book of the same title.

28 For analysis, see Loukaki (1997, 2008/2016).

Bibliography

Aristotle. *Poetics*. Available at: https://eclass.uowm.gr/modules/document/file.php/EETF189/ΑΡΙΣΤΟΤΕΛΗΣ.pdf (Accessed 28.05.2019).

Badiou, A. 2004. 'Art and Philosophy.' In *Handbook of Inaesthetics*. Stanford: Stanford University Press, 1–15.

Benjamin, W. 2006. 'On Some Motifs in Baudelaire.' In Eiland, H. and Jennings, M.W. eds., *Walter Benjamin: Selected Writings, 4: 1938–1940*. Available at: file:///C:/Users/usr1eap/Downloads/Benjamin%20on%20Baudelaire.pdf (Accessed 09.07.2019).

Bernal, M. 1987. *Black Athena The Afroasiatic Roots of Classical Civilization 1: The Fabrication of Ancient Greece 1785–1985*. London: Free Association Books and New Brunswick: Rutgers University.

Boardman, J. 1996. *Greek Art*. London: Thames & Hudson.

Braudel, F. 1994. *A History of Civilizations*. New York: The Penguin Press.

Debord, G. 1994. *The Society of the Spectacle*. Cambridge, MA: Zone Books.

Dell, C. 2010. *What Makes a Masterpiece? Encounters with Great Works of Art*. London: Thames & Hudson.

Diamond, J. 2019. *Upheaval: Turning Points for Nations in Crisis*. New York: Hachette.

Eagleton, T. 1990. *The Ideology of the Aesthetic*. Oxford: Blackwell.

Elkins, J. ed. 2006. *Art History Versus Aesthetics*. New York and London: Routledge.

Elkins, J. ed. 2007. *Is Art History Global?* New York and London: Routledge.

Foster, H. 2011. *The Art-Architecture Complex*. London: Verso.

Gero, R. 2006. 'The Border of the Aesthetic.' In Elkins, J. ed., *Art History Versus Aesthetics*. New York and London: Routledge, 3–18.

Guyer, P. 2005. 'The Origins of Modern Aesthetics: 1711–36.' In Kivy, P. ed., *The Blackwell Guide to Aesthetics*. Oxford: Blackwell, 15–44.

Harvey, D. 2000. *Spaces of Hope*. Edinburgh: Edinburgh University Press.

Harvey, D. 2006. 'Space as a Keyword.' In Castree, N. and Gregory, D. eds., *David Harvey: A Critical Reader*. New York: Wiley, 270–294.

Harvey, D. 2012. *Rebel Cities. From the Right to the City to the Urban Revolution*. London: Verso.

Hegel, G.F. 1976. 'The Philosophy of Fine Art.' In Hofstadter, A. and Kuhns, R. eds., *Philosophies of Art and Beauty. Selected Readings in Aesthetics from Plato to Heidegger*. Chicago: The University of Chicago Press, 382–445.

Herzog, J. and de Meuron, P. 2008. 'How Do Cities Differ?' In Herzog & de Meuron *1997–2001. The Complete Works*, vol. 4. Basel: Birkhäuser, 241–244.

Hofstadter, A. and Kuhns, R. eds. 1976. *Philosophies of Art and Beauty. Selected Writings in Aesthetics from Plato to Heidegger*. Chicago: The University of Chicago Press.

Horton, A. 1999. *The Films of Theo Angelopoulos: A Cinema of Contemplation*. Princeton: Princeton University Press.

Kant, I. 1952. *The Critique of Judgment*. Oxford: Clarendon Press.

Kienast, H. 1995. 'Athener Trilogie. Klassizistische Architektur und ihre Vorbilder in der Hauptstadt Griechenlands.' In *Antike Welt* 26(3), 161–176.

Lefebvre, H. 1992. *The Production of Space*. Oxford: Blackwell.

Lefkowitz, M.R. 1997. *Not Out of Africa: How Afrocentrism Became an Excuse to Teach Myth as History*. London: Basic Books.

Loukaki, A. 1997. 'Whose *Genius Loci*? Contrasting Interpretations of the Sacred Rock of the Athenian Acropolis.' In *Annals of the Association of American Geographers* 87(2), 306–329.

Loukaki, A. 2007/2009. *Mesogeiake Politistike Geographia kai Aistheteke tis Anaptyxes: He Periptose tou Rethymnou (Mediterranean Cultural Geography and Aesthetics of Development. The Case of Rethymnon, Crete)*. Athens: Kardamitsa.

Loukaki, A. 2008/2016. *Living Ruins, Value Conflicts*. London: Routledge (first edition 2008: Aldershot: Ashgate).

Loukaki, A. 2014/2016. *The Geographical Unconscious*. London: Routledge (first edition 2014: Aldershot: Ashgate).

Loukaki, A. 2019. 'Ancient Greek Space-times and Present Cyberspatialities.' Paper presented at *SpaceTime*, the 13th Conference of the Society for Science, Literature and the Arts, 25–28.06.2019, National Hellenic Research Foundation, Athens. Abstract Available at: http://st-slsaeu2019.indigoflicks.com/programme/.

Loukaki, A. 2020. 'Byzantium-Baroque: Shaping and Overcoming Aesthetic Boundaries.' In Petridis, P. and Muller, A. eds., *Mediterranean Worlds-Perceptions and Transformations of Space*. Proceedings of International Conference, National and Kapodistrian University of Athens and University of Lille 3, chapter 12. Lille: University of Lille.

Loukaki, A. and Plantzos, D. 2018. *Techne, Xoros kai Opseis Anaptyxis sten Ellada tes Krises (Art, Space and Aspects of Development in Crisis-Hit Greece)*. Athens: Leimon.

Luger, J. and Ren, J. eds. 2017. *Art and the City: Worlding the Discussion through a Critical Artscape*. London: Routledge.

Paisios, N.P. 1996. *Grammata Demetre Pikioni-N. Chatzekyriakou Ghika (Letters of Dimitris Pikions and Nikos Hadjikyriakos-Gkikas)*. Athens: Ikaros.

Plato. *Phaedrus*. Loeb Classical Library. Available at: www.loebclassics.com/view/plato_philosopher-phaedrus/1914/pb_LCL036.469.xml (Accessed 17.05.2019).

Rampley, M. 2006. 'Art History Without Aesthetics: Escaping the Legacy of Kant.' In Elkins, J. ed., *Art History Versus Aesthetics*. New York and London: Routledge, 161–166.

Rancière, J. 2009. *Aesthetics and its Discontents*. Cambridge: Polity Press.

Redfield, M. 2006. 'Island Mysteries.' In Elkins, J. ed., *Art History Versus Aesthetics*. New York and London: Routledge, 269–290.

Schaper, E. 1968. 'Aristotle's Catharsis and Aesthetic Pleasure.' In *The Philosophical Quarterly* 18(71), 131–143.

Scharfstein, B.A. 2009. *Art Without Borders-A Philosophical Exploration of Art and Humanity*. Chicago: The University of Chicago Press.

Schürmann, E. 2006. 'Art's Call for Aesthetic Theory.' In Elkins, J. ed., *Art History Versus Aesthetics*. New York and London: Routledge, 181–185.

Scruton, R. 1979. *The Aesthetics of Architecture*. Princeton: Princeton University Press.

Scully, S. 1990. *Homer and the Homeric City*. Ithaca and London: Cornell University Press.

Seferis, G. and Tsatsos, K. 1988. *Enas Dialogos gia ten Poiese. (A Dialogue on Poetry)*. Athens: Hermes.

Sejten, A.E. 2016. 'Art Fighting its Way Back to Aesthetics-Revisiting Marcel Duchamp's Fountain.' In *Journal of Art Historiography* 15, 1–8.

Stambolidis, N. and Berggruen, O. eds. 2019. *Picasso and Antiquity-Line and Clay*. Athens: Museum of Cycladic Art.

Summers, D. 2003. *Real Spaces: World Art History and the Rise of Western Modernism*. London: Phaidon.

Tarkovsky, A. 1989. *Sculpting in Time*. North Yorkshire: Combined Academic Publishers.

Tountas, Y. ed. 2016. *The Health of Greeks in Crisis*. Available at: www.dianeosis.org/wp-content/uploads/2016/03/ygeia_singles_complete_ver02.pdf (Accessed 03.09.2019).

Tziovas, D. ed. 2017. *Greece in Crisis: The Cultural Politics of Austerity*. London: Bloomsbury.

Varoufakis, Y. 2017. *Adults in the Room*. London: Bodley Head.

Wertheim, Ch. 2006. 'Why Kant Got It Right.' In Elkins, J. ed., *Art History Versus Aesthetics*. New York and London: Routledge, 172–180.

Wolf, M. 2019. *Greek Economy Shows Promising Signs of Growth*. Available at: www.ft.com/content/b42ee1ac-4a27-11e9-bde6-79eaea5acb64 (Accessed 20.05.2019).

Part 1

Creating, imitating, and destroying the urban sublime and its materiality

From Athenian classicism to neoclassicism and the present

1   Ancient Greek cities as works of art

Manolis Korres

The emergence of the Greek polis and its public amenities

With the transition from the hunting-gathering to the farming phase of human life, the first notable permanent settlements were formed, which very soon developed in size and building quality resulting in the creation of small towns and then large cities. The earliest known specimens of the phenomenon are found in the Middle East, mainly in areas with rivers and alluvial soils. In such places, with no difficult terrain, the shape of the city could effortlessly correspond to a program of function and symbolism.

When over time this development also covered Greece, what was most influential on the shape and geographical distribution of the cities was the strong relief of a land consisting of small valleys and plains between high mountains, which for each city constituted physical and ultimately administrative boundaries. However, since in Greece, as in other similar countries where the mountains are adjacent to the sea, the existence of peninsulas, capes, bays, inlets, promontories, and coastal islands is the rule, the most important cities – craft and commerce centers, since their first appearance before four millennia – were built by the sea, or a short distance from it. In Mycenaean Greece, cities based on agriculture and commerce were surrounded by capable agricultural areas, had a high, strategically built citadel, and were connected by a highly refined road network.

After centuries of decline (Dark Ages following a warfare and migration period, or simply because of a widespread socio-economic crisis), old *Mycenaean* cities, or others on new sites, are beginning to flourish again in the 8th century BC, based again on the same economic terms of agriculture, craft and commerce, but with a new administrative system, that of the *Polis*, the self-governing unity of an urban center and of its rural country.

This new urban development has the following common physical characteristics: dynamic central, bi-central, or axial development adjusted to the features of the terrain; an initially elementary fortification; and a direct connection to at least one port.

These cities had also common ideological characteristics: an awareness of a common historical identity based on the use of the same language and the worship of the same gods, but at the same time insistence on constant territorial, often warring claims, which sometimes seemed to be the purpose of their existence. To serve this purpose, the Cities devized, preserved, and presented their particular mythological-historical narratives, which at the same time contained links to Pan-*Hellenic* mythical and historical traditions. The focal point of the city's ideological background was its founding by an ancient leader, the ancestor of all citizens, whom they called the city's *Heros Ktistes*, the Founding Hero. In larger Poleis, such as Athens, which came from the political union of smaller Demoi (communities with their own townships), besides the common Hero, such as Theseus in Athens, there were other, sometimes older, local Heroes. They connected with each particular Demos and were also venerated by citizens descending from them.

A general principle in all Greek cities, as in most of the rest of the ancient world, was the prohibition of burials within the city. The cemeteries had to be outside of it, but with one exception: the tomb of the city's founder had to be inside the settlement, near the supposed first house, where he had lived before the city was founded. The great antiquity of this tomb was for the citizens a way to prove their exclusive right to own their land.

At a time when there was no photography, ID card, or other modern means as evidence of identity, the only way to satisfy this need was to witness to fellow citizens. Thus, everyone was obliged to participate in common activities that took place in public spaces to ensure the recognition of individual identity and the right to the privileges deriving from the united power of all. In this way the action of individuals was public rather than private. This was self-evident, and if for Aristotle it happens to be an important issue, it is only because in his days (4th century BC) the increasing turn of citizens to private interests threatened public life.

Public spaces, absolutely necessary by the nature of city life, were places of worship, places of commerce and gymnasia, places of military exercize, and important city holidays. The general participation of citizens determined the extent of the sanctuaries, the market, the gymnasia, and the characteristics of the central streets, with the result that Greek cities had unusually large public spaces. Places of worship, including the Founder's Heroon, generally developed at sites where there were mythical or archaeological evidence linking them to the ancient (Mycenaean) *Times of the Heroes*. Trade places were developed inside the city where important local and interurban roads converged. Finally, gymnasia were set up and developed on smooth ground just outside the city, exactly where in the event of a war the citizens would have to join to defend it.

As a rule, gymnasia were ideologically linked to the city's historical traditions through the worship of a particular Hero and through ceremonies dedicated to him. Worship's programs as a basic means to create and preserve historical identity and social coherence included sacrifices, other symbolic performances, processions, and competitions, primarily musical and athletic. Centuries later, with the development of the theater, stage plays, drama, and comedy were added to the city's performances and competition programs.

Naturally, the general context of citizens' lives was highly public; therefore the use of the arts was likewise public. This allowed the production of luxurious artworks in large numbers and the development of rules made to serve the highest artistic ends and express the state's (*Polis*) ethical and aesthetic ideals. These projects were financed not only by the state but also by individuals, under the condition that there was an open process of proposals, discussion, and approval of the type and place of any monument in the public space (more about the place of art in the city is said later in this chapter).

Due to the strong geophysical division of the country, both land and sea, its inhabited part was highly fragmented and the division of the Greeks into communities, cities, and major ethnic groups, such as Ionians, Aeolians, Dorians, Thessalians, and so forth, was also intense, with special features and special traditions. However, not only did all the cities have the same basic institutions as cults and agonistic programs (athletics, music, and stage), but they also preserved the memory of their common origin and common history. Not accidentally this is imprinted in the *Iliad*, as late as the 8th century BC, where various names (e.g., *Danaoi*) denote all Greeks; in the same way the name *Hellenes* will later (beginning of the 5th century BC) become the official one. In this context of consciousness of common origin, the sanctuaries of *Olympia, Delphi*, and *Isthmia*, which were originally local in scope, became of particular importance to all Greeks, and the festivities in which the competitions (athletic, etc.) were a major component, became Pan-Hellenic.

The emergence of the aforementioned new type of city (the *Polis*) in Greece as a major historical phenomenon coincides with an unprecedented growth in population; replacement of the older writing system (*Linear B*) with a much higher (?) one (Greek alphabet as a major development of the *Phoenician*); writing of the Homeric epics; Greek colonization, that is, the founding of overseas Greek cities across the Mediterranean with a particular density in southern Italy and Sicily; and the establishment of the Pan-Hellenic festivals and competitions, beginning with Olympia, already before the middle of the 8th century BC. These phenomena are so interconnected that it is not possible to discern which precede as causes. It seems, however, that the need to preserve historical bonds was realized when the diaspora created? by colonization threatened their strength. It was then that these bonds were strengthened by the Pan-Hellenic prevalence of epic works such as Homer's and competitional festivals such as Olympia's. But while the constituent type of cities, as we described it, was about the same (sanctuaries, markets, gymnasia), local territorial and other conditions, intense as a rule and varied, resulted in a multitude of different urban arrangements and associations with the broader inhabited space whose systematic analysis and classification is a real challenge for the modern scholar.

Greek cities had at least one citadel and other fortifications along a perimeter line, which, as it followed naturally more favorable defensive places, was irregular, with gates where it encountered pre-existing roads that converged toward the center of the city. These first roads in their turn determined the location of the central and other sanctuaries, as well as the market and public buildings. A series of smaller roads divided the remainder into irregular building insulae. The city was dynamically developing along its main axes, and its plan, without being the product of a designer, was as good as possible to meet the functional needs of urban life.

While most of the old cities were irregular in plan, developing dynamically and functionally, the new cities in the colonies were initially designed on the base of a well-planned rectangular network of roads and streets. This system was not new in itself. It has long been used in many parts of Asia, as well as in Egypt, with well-known examples the settlements for workers built near the Pyramids 45 centuries ago. Therefore, the new in the case of the new Greek cities with a sizable regular orthogonal system (beginning in the 7th century BC), for which the term *Hippodameion* is coined (after the city-planning theorist Hippodamos, early 5th century BC), was the scientifically elaborated way of organizing the functions and spaces within it. In addition, a basic function, that of defense in the event of a siege, also observed in the new cities the same principles as in the old cities, and the city walls followed an irregular outline, little influenced by the orthogonal plan within it.

In terms of the history of city planning principles, the Greek cities, even the designed ones, do *not* conform to the ideas of plan symbolism supposed to have been occasionally applied by town planners in the past. These ideas relate to cities of other cultures, with regular rectangles, squares, polygons, or circular shapes; the use of significant or symbolic numbers (quantities of parts and various dimensions, overall and particular); and in a regular orientation consistent with the cardinal points. City plans with such characteristics were supposed to be images of a divine cosmic order, incorporating invisible magical powers necessary to prevent a military attack or malicious spirits. In fact, cities exactly square or circular, precisely oriented to the cardinal points were few and yet only in Asia. The Greek cities, as well as the Roman cities, even those of later periods that evolved from large square walled camps, rarely conform with the cardinal points, but instead their orientation seems better to ease flow of water into the sewers and down the drains and according to Vitruvius better to cope with the site's prevailing winds, with the aim of a balanced and effective ventilation. However, ancient thought (as well as that of some in our times) often lashed out at symbolic interpretations and by no means accidentally at the level of mythical intake. Even Rome, as irregular as it happened to be, was imagined as a perfect square, *Roma quadrata*, and alternatively in some medieval imagery as a perfect circle! But while an initial square could be supposed on the quadrilateral top of the Palatine Hill, the circle is completely absent from any phase of the *Eternal City*. In Europe since the time of the Italian Renaissance, 'Ideal Cities' or related spatial arrangements, precisely circular or regular polygonal, always scarce and rather small, have been the result of Platonic-type design principles combined with functional theories including (among other issues) the uniform application of ballistic principles for all points along a city wall. In the latter, as in other similar modern cases of perfect circular, square, or other symmetrical layouts, rationality has very often been combined with artistic intentions and monumental pursuits, so that the city is not only a product of practical rationality but also a work of art.[1]

The ancient city as a work of art

At this point, with the help of the following, it is appropriate to define exactly what it means to recognize a city as a work of art. An art museum, for example, considered as an edifice, may be a work of art, only if itself has an appropriate architectural fashioning and not merely because it houses art masterpieces. Similarly cities, although pre-eminent places where art is being born and experienced, can be works of art (by themselves) only when they are designed with artistic intentions. That is only when simultaneously the following occur:

1   Beyond realistic traffic, military, economic, or other practical prerequisites, a city plan, entirely or in its more substantial parts, follows idealistic geometrical forms and patterns, which, depending on circumstances, could possibly comply with the physical and practical conditions, while often defy or collide with them. Such geometric arrangements, being made for visual and more importantly for mentally perceptible beauty, are axial symmetry (a1), central or rotational symmetry (a2), simple numerical proportions (a3), repetition of standard shapes (a4), etc.

2   Many buildings or landscaping operations, that is secular public buildings (b1), sacral buildings (b2), market-places (b3), assembly-places (b4), terraces and platforms (b5), votive, honorary, and commemorative monuments (b6), fountains (b7), bridges (b8), fortifications (b9), and so forth, beyond their utilitarian purpose, which sometimes can be quite limited, are always artistically important. They are conceived as self-sufficient units of any acceptable building type (c1, c2, c3. . .) and architectural style (d1, d2, d3. . .) and likewise as necessary components of a composition in urban scale.

3   The arrangement of these buildings and artistically fashioned open spaces, most commonly being linear (e1), curvilinear (e2), circular (e3), 'free' (e4), juxtaposed (e5), dynamically balanced (e6), and so forth, is directly linked to the city plan. In some rare cases of ancient cities, this plan encompasses their whole expanse, while in most cases it is limited to one or more well-selected parts (certain sanctuaries, agoras, colonnaded streets, temple platforms, etc.). Usually, such concepts are being realized in steps that very often remain unfinished. Nevertheless, despite their additive and fragmentary character, urban plans bear witness of the underlying concepts and their artistic intensions.

Obviously, the possible combinations of these elements (a1, a2. . ., b1, b2. . ., c1, c2. . ., d1, d2. . . e1, e2. . .) are quite numerous, and since they are associated with the various sizes of cities (f1, f2, f3 = small, medium, large), the various types of terrain (g1, g2, g3, . . .), and the eras of history (Archaic, Classical, Hellenistic, Roman), with their specific ideas and circumstances (political, economic, artistic, etc.), they easily result in a vast number of particular cases that invite us to document, classify, and analyze their data in order better to understand the processes that gradually transformed so many ancient cities into large composite works of art. A common property of these cities is a certain theatricality (not to be confused with the same term's meaning in our theory of contemporary art and architecture).

In addition to this highly artistic theatricality, the outer figural decoration of various buildings and in particular of those belonging to the aforementioned categories b6 and b7 transformed many ancient cities into vast outdoor art exhibitions, of which the current archaeological museums, even the richest, are not but faint reflections.

The large axial compositions of building blocks and the pompous streets with colonnades, peristyles, and so forth by which cities acquired the character of artworks, while simultaneously, thanks to the sculptural decoration of their public space, being museum-like, had been progressively formed until the Hellenistic times. However, in earlier times and sometimes in classical times (like in Sparta as commented by Thucidides) these elements were scarce and sometimes absent.

At the time of the founding of the Greek City (8th century BC), institutions and functions were the dominant elements. Not the public buildings that, if any, were completely humble. It is therefore clear that, at this stage of development, function preceded its architectural fulfilment. At that time there was no city worthy of being recognized as a work of art. In time, however, some cities developed their architecture as high art, first through the erection of temples that within a few decades in the early 6th century acquired gigantic dimensions, followed by the construction of *stoas* and gate buildings, and then the creation of other public buildings. At this stage, architecture was worthy of the institutions it housed and of the spiritual goals it served. This course remained unchanged until the 5th century.

In the 4th century, notwithstanding three or four exceptions, giant temples were no longer built; however, other types of buildings were being developed, such as the *gymnasion* as a large complex, the Bouleuterion as a specialized administrative building, and the theater as a building of major importance to city life. In this stage of city development, architectural forms no longer have the aesthetic power of those of the mature archaic era, and structural technology is constantly looking for easier (and less expensive) ways of executing works. Stone carving is no longer as perfect as in archaic and early classical works, but now more than ever before, the composition of different spaces and solid volumes in one edifice and the synthesis of more buildings and spaces in one unit is achieved.

The importance of the architectural façade
and of the monumental entrance

The architectural development goes with a delicate process of outstanding importance: the consideration of the architecturally shaped surface, as a self-sufficient object, independent of the rest of the building organism. In fact, this process, so new to classical Greece, was already well known in world architecture (Egypt), as it was also known in Mycenaean Greece (highly monumental façades of vaulted tombs). In all event, to decide the autonomy of the stylistically formulated surface implies the idea that, although this surface results from the process of assembling distinctive structural elements to create a three-dimensional object, it may also be regarded as an immaterial 'garment' independent of the object or the creative process, suitable for the 'wear' of other bodies, not necessarily of the same structure or use. A typical case of such treatment, which may have contributed to the maturation of the 'garment' idea, is well known: structures created by assembly (machine-like fitting together of parts) were being imitated in products made out of a single continuous mass by a removal (subtraction) procedure (carving, cutting, etc.), or by casting (plaster, terracotta, concrete, etc.).[2] The process is described today by *Bekleidungstheorie* (Gottfried Semper, *Der Stil*, 1860), but it seems that its essence has been an artistic issue since antiquity.

As the available evidence shows, the autonomy of the 'garment' that had been somewhat restrained by the 6th century BC (e.g., triglyphs on all sides, but also inside a temple) evolved very rapidly, and soon semi-columns began to appear on the outside and inside of buildings.[3] The next step involved the colonnade in its aspect as a major element enveloping a building, so that this relation was reversed and the colonnade enveloped a space, roofed (Parthenon, 447 BC) or unroofed (Pompeion, 420 BC). This was nothing new to the typology of the colonnaded architecture. Many centuries before it had taken place on a highly monumental scale in Egypt and in a simple form in Mycenaean Tiryns. But in the case of Classical Greek architecture, its achievement had to face some prohibitive terms posed by building typology and rules of style.

Given the special conditions in the evolution of classical Greek architecture, the Propylaea, built in the years 437–432 BC on the Athenian Acropolis constitute the most important phase of it in terms of both composition and flexibility in the use of colonnades. The entire composition, 65 meters wide, is made up of a central, highly monumental hexastyle entrance building, the Propylaea proper, with an axis parallel to the Parthenon, two large wings perpendicular to its sidewalls and two smaller wings perpendicular to the west side of the latter. These five parts replace what used to be there before: five different architectures, only one of which was a gate building, while the other belonged to four different sanctuaries, two outside and two within the Acropolis. The five pre-existent buildings, placed close to each other, differed in height, orientation, style, and dimensions, and the only means to replace them with other larger ones was the unifying composition conceived by architect Mnesikles.

Another achievement of the architect was to make the façade of the central building and the façades of the small wings in a Π-shaped arrangement so that the two end columns of the latter stood in alinement with the six much larger columns of the main building. In this way the courtyard of the Propylaea was open on one side and colonnaded on the other three. Thus, for the first time in Classical architecture the colonnades of a composite building nearly surrounded an outdoor space. The composition included a huge 22-meter-wide axial ramp from the foot of the hill to the Propylaea, and gates on either side of it to the adjacent sanctuaries, including that of Athena Nike, whose tower-shaped platform was aligned with the ramp. In this way, due to the impressiveness of the building's axiality – and despite the steepness of the terrain – a much greater area, one hundred meters outside and inside the gates, was conceptually transformed to an integral part of the new building's spatial composition and domain. Thus, a big step was made toward a wider concept encompassing buildings and their surroundings as an artfully designed entity on a nearly urban scale. Again, this type of design was not quite new. Suffice to analyze the architectural composition of Persepolis, preceding Propylaea by nearly eighty years. Still, since then progress continued, there is more of this quality in the Athenian example. By designing the Propylaea, Mnesikles laid down the still valid rules for any large axial composition with symmetrical building volumes and monumental access. The examples are innumerable and surely readers will have many of them in their memory.

But as the Propylaea, like their predecessor, a propylon with columnar façades, stood in the place of the once fortified Wall Gate of Mycenaean times, the occasion seems quite relevant for some observations on the use of architectural and artistic fashioning in fortifications as a symbolic embellishment. That fortifications are generally roughly built structures with no artistic intention is a well-known fact. Exceptions are therefore highly interesting. On the North Aegean island of Thasos, the inside façade of the principal southern city gate, that of 'Zeus and Hera,' was in early 3rd century BC architecturally fashioned with pilasters bearing a Doric entablature and an upper gabled tier with Ionic columns and entablature as well. In the same and other city gates there were sculptural representations dating back to ca. 300 BC (Zeus, Hera) or to about 500 BC (Dionysos, Hermes, Herakles, Silen). A somewhat different symbolism, but always of excellent artistic quality, characterizes the Lion Gate at Mycenae, made one thousand years before the Zeus and Hera Gate of Thasos.

Of about the same old age and same symbolism with the Mycenean achievement is the Lion Gate of the Hittite capital city Hattusa, a masterpiece of shear Cyclopean masonry and fine monumental sculpture. The use of Lions or mythological beings as guardians of gates, artfully executed in hard stones of monumental size, was continued with excellent results in the next millennium (Zinjirli, Tell Halaf, Dur Sharrukin, Persepolis). In most cases the whole setting included large numbers of accompanying sculptural representations and decorative elements, transforming the city walls, their courtyards and the relevant platforms and ramparts into grandiose works of art strongly dominating the image of the city.

The walls of classical Athens and the urban topography

Particularly interesting are the cases of reusing sculptures and architectural members in the walls of Athens, built immediately after the Persians were expelled (479 BC). In the wall section passing through Kerameikos,[4] stones from burial monuments were used roughly re-carved without any artistic intention. In contrast to this, the enormous column drums of the Olympieion, at the city's eastern outskirts, were carefully bisected and redressed to form the lower part of the nearby city gate and its towers, with limited artistic treatment.

At the northern wall of the Acropolis, massive architraves, triglyphs, metopes and cornices of the temple of *Athena Polias* were reused without any re-carving. They were placed intact, precisely in the order they originally had in the temple, so that the result exactly likened the part of the temple that was previously visible from the agora. In the prevailing view, the Athenians thus wanted to create a monument commemorating the disaster of the war. Apart from this, the anticipated rebuilding of the Parthenon, but probably not of the temple of Athena Polias, may have added the purpose of using the wall as a museum preserving as much of the temple's architecture as possible.

Shortly after the intensive completion of the fortification of Athens, the construction of the Long Walls from Athens to the two ports of the city began. In total, two long walls were built all along the way to Piraeus, forming a nearly 200 meter wide and 6 kilometer long corridor and one to Phaleron, 4 kilometers long. It is very interesting to note that over many kilometers the corridor to Piraeus was situated on very low flat terrain, and therefore its position could be freely defined on the base of principles independent of intermediate obstacles. On this basis the street connecting the city with the harbor was defined by a straight line passing through the middle of the Acropolis Rock and the middle of the mouth of the central port, nine kilometers to the west. This line crossed the fortification of Piraeus at the point where the gate facing Athens was built. Similarly, when much later another wall facing Piraeus, the so-called *Diatechisma*, was added close to the Akropolis, and since the terrain was not prohibiting, the new gate was defined by the same line. Anyone who was looking from the Acropolis to Piraeus, saw on one line the Piraeus Street with at least two gates and the far distant mouth of the main harbor. Again, anyone nearing the harbor by sea, or walking along the street to Athens had the unique experience of heading straight to the Akropolis' top. Today, since over a considerable part of its length the present Peiraic street nearly coincide with its predecessor, the experience remains partially the same.

However, the magnificent visibility of the Acropolis from Piraeus Street is subject to a substantial restriction, due to the presence of the western hills, which, as one approaches Athens, gradually conceal the Acropolis until after the middle of the route. There, the altitude is only 13 meters, but the ridge of hills, with a minimum height of 99 meters, completely conceals even the Parthenon.

By no coincidence, after this point, the current Piraeus Street with a well perceivable left turn takes a northerly direction so that, bypassing the hills, it passes the Kerameikos at an altitude of 48 meters and ends at *Omonoia Square* (altitude 70m). A similar road bypassing the hills, but closer to them, existed earlier, but the main road proceeded with slight bends through the valley of the hills heading toward their saddle (altitude 99m), where the gate of the Diateichisma was placed, and after a distance of nearly two hundred meters downhill it smoothly reached the meeting point of the roads to Pnyx, Agora, Acropolis, and East Athens. Given the said bends whose shape is dictated by the relief of the hills, the far west of the city straight line section of the ancient (and the present street) could have any direction up to 2 degrees north or south of the one chosen with no effect to its overall length. Therefore, the selection of a particular line aiming precisely at the Acropolis from such a great distance is solely understandable as an artistic act.

Similar alignments are also observed on other streets or areas of the city. The axis of the Demosion Sema (a 40 meter wide street stretching from the Dipylon[5] to the Academy[6]) points to the Propylaea, while the axis of Dipylon points to the Parthenon. In both cases, however, the result would be truly convincing if the axes in question were turned by one degree clockwise. But in such matters of defining axes and of their significant or symbolic purposes, one degree of deviation is not a negligible amount, and therefore the significance of these axes' direction has still to be debated.

The layout and orientation of classical Athenian monuments

Related to the foregoing is the theory that the lay-out of the Akropolis, as well as of other sites, as an ensemble of buildings and accompanying features follows rules of visibility from important points within a set of well-balanced angles of view (the Doxiades theory). This theory obviously extends the well-known thesis that classical architecture intentionally avoids frontal and prefers oblique views. Concerning the latter, the unbiased observer or reader will state that if one situation is a common case and another is its exception, then in the topic under consideration the oblique view was and still is the common and the frontal is the exception.[7] With the passage of time, thanks to its uniqueness, axial access has been used on a number of occasions, especially in Asia and Egypt, to emphasize the importance of some buildings and their setting. At any rate, the absence of axial access continued to be the common situation, even in Asia and far more in Greece. Therefore, the oblique access to and the over-angle viewing of a Greek temple should be considered not as the result of a certain theory but rather as a given situation.

As for the theory of visibility from important points within a set of well-balanced angles of view, it is sufficient to say that the present position of the Parthenon (Parthenon III) and its orientation (making its viewing from the Akropolis entrance unavoidably oblique), had been determined by its archaic predecessor in a slightly different orientation, 120 years before Perikles, and in the current orientation, by Parthenon I, 60 years earlier. In the time of Parthenon I, the Propylon was oriented very differently from the current Propylaea and thus the position of the Parthenon with respect to the Propylaea was not the result of a single design, but rather a consequence of several decisions made by different specialists and administrators over a period of many generations. Still, as an ongoing study shows, certain angles of view, combined with observable alignments, constitute a new evidence of ancient site design.

Another issue for which there is more convincing evidence is that of *parallelism*, the best example of which, that of the Propylaea to the Parthenon, has already been mentioned. Other noteworthy cases are the geometric parallelism of precincts and buildings in the *Asklepieion* and its surroundings in the Southern wall of the Acropolis, and the parallelism of all buildings at the west side of the ancient Agora. With the emergence of long stoas, the principle of parallelism was extended to that of *perpendicularity*; the best and most well-analyzed example is that between the Middle Stoa and the Stoa of Attalos in the Athenian agora.

Generally, the orientation of most ancient Greek buildings, as being the result of other considerations, greatly deviates from the cardinal points and therefore when this is not so it is worthy of special mention. This is the case of the *Olympieion*: the city's largest temple is in complete agreement with the cardinal points. This should not be attributed to a late influence 'from the Orient,' say from the time of Antiochus II (who in the early 2nd century BC financed the temple's construction), because the crepis, the three-stepped base of the temple, is still that of the original Doric temple, made in the 6th century BC. Therefore, the astronomical orientation of the temple is a genuine concept of the Archaic period. And there is more: on the other side of Ilissos River, at least two buildings were made parallel to the Olympieion. One of them is the so-called Temple on Ilissos, published by Stuart and Revett in 1762.[8] The other is one whose foundations were discovered four years ago, a short distance to the east of the first.

Classical theaters

In the 5th century BC, when the great classics composed their great works, theaters with permanent site and monumental form still did not exist. An elementary orchestra, simple wooden benches and a wooden scenic structure at the other end were the only physical features. This minimalistic situation was similar to that of the Agora, which at the high moments of its own history had not any remarkable architectural design.

In the following century the theatrical institution was consolidated. Temporary structures were replaced by permanent stone ones of monumental scale and sophisticated design: the *orchestra*, originally quadrilateral, was transformed into a circle, and the area provided for the spectators, the *cavea* (Greek: *koilon*), became approximately semicircular. In about the same time, a similar transformation, likewise a creative response of designers to the theorized equality of all spectators, took place in planning and fashioning stadia: one of their originally rectangular ends became semicircular.

Still, the new monumental components of the theater, the scenic building included, maintained their original spatial relationship: the shape and size of each one did not depend on the type and size, nor the exact distance of the others.

In the next two centuries only minor but sequential adjustments were made: the stage building is mutated at both ends with the addition of *paraskenia*, two wings protruding toward the *cavea*, and of screen-like gates in the space left between the *paraskenia* and *cavea*. This development is similar to that of the agora, where stoas and other buildings, screen-like passages, and more were added all round, not always for purely utilitarian purposes, but in certain cases for their usefulness as architectural components of a continuous spatial definition.

The next transformation of the theater as a building type took place in Roman times and aimed to integrate its heterogeneous functional and structural components into a single entity of uniform height and architectural style, defined externally and internally by rhythmically repeated architectural themes, including columns, pilasters, entablatures, cornices, niches, and arched passages.

Once again, the architectural development of the theater has been similar to the corresponding Hellenistic and Roman evolution of the agora into a single peristyle, or to the emergence of the Roman imperial forum late in the 1st century BC.

The architectural development of theaters, as observed over a long period of time, was strongly dictated by operational reasons, the changing stage, and other issues of the performances, but these reasons alone explain only a part of the procedure. The other part must simply be explained as the result of a very strong and persistent desire to transform urban space into an artful architectural landscape. This desire would, of course, not be fulfilled if it did not appear to be so well combined with the political, social, and economic feasibility of large public works. Such major public works in Athens were the Olympieion from the mid-4th century on, the two-tiered Stoa of Eumenes (164m long, ca. 180–170 BC), the aforementioned two stoas in the agora, a nearly 200 meter long stoa to the east of the Tower of the Winds, two more two-tiered stoas nearby, the large edifice of the so-called Roman Agora, with its spacious peristyle, the long colonnaded street to the east of the Tower of the Winds, the Library of Hadrian, the Basilica at the north side of the agora, the so-called Pantheon to the north of the Tower of the Winds, and the Odeum of Herodes Atticus.

In conclusion, one might say that great ancient cities such as Athens, Corinth, Ephesus or Miletus and many other smaller ones were not only places of an overproduction of works of spirit and fine arts, but were themselves artful compositions of public spaces and luxurious buildings. However, one can also observe that the era of great urban development and beautification was no longer the time of the greatest spiritual, moral, or political standards. These virtues belonged to the maturing time of few generations after the emergence of the *polis* as a politically autonomous entity.

Notes

1 For more details, see Eaton (2001).
2 For instance, massive vessels or furniture with details mimicking assembly of structural parts, monumental rock-cut tombs in imitation of free-standing buildings assembled with distinctive architectural members, etc.
3 See the large temple in Akragas (present-day Agrigento), Sicily, early 5th century BC.
4 The First Cemetery of Athens, to the north-western end of the city.
5 The Dipylon (Greek: Δίπυλον, 'Two-Gated' entrance) was the main gate in the city wall of Classical Athens, located in Kerameikos.
6 The meeting location of Plato's Academy was originally an olive grove in the north-western outskirts of Athens. Plato acquired property about 387 BC and used to teach. At the site there had also been a park and a gymnasium sacred to the legendary Attic hero Academus (or Hecademus).

7 For example, a person or animal moving toward an isolated pair of trees in a meadow, for instance seeking for fruits, most probably chooses the shortest or easiest way, which most probably forms a random angle with the line defined by the two trees and it happens quite rarely that this way coincides with the perpendicular bisector of the line defined by the trees – in short the axis of their symmetry. Still more exceptional will be the case that although the starting point is quite off from this axis of symmetry, the visitor of the trees chooses a detour to reach first this axis, and then to move along its length toward the trees. This can only happen when the visitor is a person and not an animal, and whenever it happens the motivation is something more than just the desire for the fruits. It could be the idea of equal distances, a curiosity for visual experience, or finally the discovery that among the countless possible routes leading to a pair of points, there is a unique one, their axis of symmetry.

8 https://www.ascsa.edu.gr/uploads/media/hesperia/147912.pdf (Accessed 22.02.2020).

Bibliography

Benevolo, L. 1976. *Storia della Città*. Bari: Laterza.

Benevolo, L. and Albrecht, B. 2002. *Le Origini dell'Architettura*. Bari: Laterza.

Brownlee, D.B. 1989/2017. *Building the City Beautiful*. Philadelphia: Philadelphia Museum of Art.

Coulton, J.J. 1976. *The Architectural Development of the Greek Stoa*. Oxford: Clarendon Press.

Eaton, R. 2001. *Cités Idéales. L'Utopisme et l'Environment (Non) Bâti*. Antwerp: Fonds Mercator.

Glotz, G. 1968. *La Cité Grecque*. Paris: Albin Michel.

Greco, E. ed. 2010. *Topografia di Atene*. Athens: Scuola Archeologica Italiana di Atene.

Gros, P. Vol. 1, Paris 1994, Vol. 2, Paris 2001. *L'Architecture Romaine du Début du IIIe Siècle av. J.-C. à la Fin de la République Romaine I*. Paris: Piccard.

Habicht, C. 1995. *Athen: Die Geschichte der Stadt in Hellenistischer Zeit*. Munich: Beck.

Hansen, M.H. ed. 1997. *The Polis as an Urban Centre as a Political Community*. Copenhagen: The Royal Academy of Sciences and Letters.

Höpfner, W. ed. 1999. *Geschichte des Wohnens*. Stuttgart: Deutsche Verlag-Anstalt.

Höpfner, W. and Schwandner, L. 1986/1994. *Haus und Stadt im Klassischen Griechenland*. Munich: Deutscher Kunstverlag.

Judeich, W. 1931. *Topographie von Athen*. Munich: C.H. Beck.

Krischen, F. 1938. *Die Griechische Stadt*. Berlin: Mann.

Kron, U. 1975. *Die Zehn Attischen Phylenheroen. Geschichte, Mythos, Kult und Darstellungen*. Berlin: Deutsches Archäologisches Institut.

Kruft, H.W. 1989. *Städte in Utopia: Die Idealstadt vom 15. bis zum 18. Jahrhundert zwischen Staatsutopie und Wirklichkeit*. Munich: C.H. Beck.

Manidaki, V. 2018. 'Theories About the Oblique View in the Study of Ancient Classical Architecture.' In Korres, M., Mamaloukos, S. and Zampas, K. eds., *Heros Ktistes (Hero Builder)*. Athens: Melissa, 125–136.

Martin, R. 1956/1974. *L'Urbanisme dans la Grèce Antique*. Paris: Picard.

Mertens, D. 2006. *Städte und Bauten der Westgriechen, von der Kolonisationszeit bis zur Krise um 400 vor Christus*. Munich: Hirmer.

Polignac, F. de. 1984. *La Naissance de la Cité Grecque. Cultes, Éspace, et Société, VIIIe-VIIe Siècles avant J.-C.* Paris: Éditions de la Découverte.

Sielhorst, B. 2015. *Hellenistische Agorai, Gestaltung, Rezeption und Semantik eines Urbanen Raumes*. De Gruyter: Berlin.

Travlos, J. 1970. *Pictorial Dictionary of Ancient Athens*. Tübingen and London: Praeger.

Ward-Perkins, J.B. 1974. *Cities of Ancient Greece and Italy: Planning in Classical Antiquity*. New York: Braziller.

2 Assaulting the archetypes

Urban materiality and the current adulation and hatred for marble in Athens

Argyro Loukaki

In Plato's *Symposium*, a woman philosopher, Diotima, invokes form as the object of human Eros. This Greek desire, affection for crystal-precise and beautiful *morphe*, balances out social entropy and chaos, promising eternity and immortality, as it merges natural potentiality with human thought which, ironically, is moulded by mortality according to Socrates. *Morphe* has received fabulous expressions in classical art and architecture, primarily that of Athens. 'We Athenians are lovers of the beautiful, yet simple in our tastes,' said Pericles.[1] Seating on the Sacred Rock of the Acropolis, heart and soul of the city, are the 'erotic' standards of Greek and Western civilization and art, the Parthenon first and foremost, with its extraordinary sculptures made by Phidias, the Erechtheion, and the Temple of Athena Nike, as well as the monumental entrance, the Propylaea, all made of white marble after the Persian Wars (see Chapter 1). Marble has become a major symbol of Greece, synonymous with beauty, veneration of the human form, democracy, freedom, and dignity, as well as a symbol of national and urban permanence and regeneration.

Marble for this great 5th-century BC Periclean building programme was extracted from the effectively organized quarries of the pyramidal Mount Pentelikon, some seventeen kilometres to the north of the city (Figs. 2.1, 2.2). Attica is abundantly blessed with building stones, as the ancient historian Xenophon attested. The Athenians initially imported *lychnites*, the highly translucent, pure, large-grained marble of the Cycladic island of Paros, but thereafter used mainly the finer-grained, also translucent Pentelic which honeys chromatically over time (Washington 1898), as literature admiringly records (see de Lillo 1992, 330). This judicious use was destined almost exclusively for edifices housing some of the highest institutions of Athens (see Chapter 1). The globally famous Pentelic marble is available in massive beds whence came the blocks for the Acropolis architectural members;[2] thus, on the Periclean Acropolis walls, foundations, sculptures, and columns grow from the native Attic soil.[3]

Identified with goddess Athena and Athens since deep antiquity, the Acropolis Rock, an everlasting pledge of collectivity, has raised Athens after many local and national catastrophes. This feat is possible only to the most essential urban monuments, knots of collective memory and urban permanence.[4] The Acropolis Rock is a place of unique experience, spatial, mythological, spiritual, architectonic, and

Figure 2.1 The Penteli quarry. Drawing restoration by Manolis Korres.

© Drawing: Manolis Korres. Published in M. Korres 1994. *From Pentelikon to the Parthenon*. Athens: Melissa, 33.

sculptural. The awe-inspiring perfection of the Parthenon, transcendent and yet made by human hands, guarantees the future of the city and the nation. But this is also a locus of freedom: not only because it incessantly launches imaginary visual flights over the Attic basin, the Saronic Gulf with its islands, and the Peloponnese (Figure I.5), but also because it validates the sheer physicality and humanity permeating classical architecture. Much later, Pentelic marble was used for the construction of neoclassical as well as of modern Athens.

In Mediterranean urban and cultural geographies, rich deposits of the natural and cultural subsoil are vitally important. If the porosity of tuff, sponge-like volcanic rock defines the character of another Mediterranean city, Naples, as a city rising from volcanic fire, its own womb,[5] Greece in general and Athens in particular are places of essentials like light, salt and marble.[6] Greek archetypes trigger the spatial unconscious and stimulate contemporary creators, foreign and

Figure 2.2 The entrance of the cave of Penteli with the 11th-century Byzantine chapels of
St. Spyridon and St. Nicholas.

© Photo: Wikipedia Commons. Source: https://en.wikipedia.org/wiki/File:Penteli_ntavelis_cave_b.
jpg#/media/File:Penteli_ntavelis_cave_b.jpg

local, like, respectively, Allen Ginsberg and James Merrill, or Giorgos Seferis and
Yannis Ritsos.[7] Ancient marbles shining in the sun against the sea, or enclosed in
museums, recapitulate the prosperity of Greece in poets Sikelianos, Seferis and
Elytis, are commentaries on the close bonds between nature and culture as well
as on the extent of landscape transformation, but also epitomize the national debt
of the present to the past. To Ritsos, marble is symbol of national freedom and
dignity:

> These marbles here don't get any bad rust
> nor do chains grip the foot of the Greek or the wind.[8]

To Seferis, marble denotes national elation after critical times:

> A little more and we will see the almond trees blooming
> the marbles sparkling in the sun
> the sea waving
> A little more and we will raise a little higher.[9]

Marble architecture in particular and urban monumental space in general has been explored from many immaterial and organizational viewpoints beside the symbolic and poetic, including aesthetic, spatial, administrative, identity-related, and material: archaeological, architectonic, planning, and technical. Urban materiality has recently emerged as an engine of possibility, enabling or disabling as it does certain actions, identities and practices which are specific to each and every city. The making of urban space as a material practice has been consistently pursued by geographers including David Harvey, Erik Swyngedouw, Neil Smith, and more,[10] while others, having taken a distance from materiality during the 1990s, have later negotiated their rapprochement (Latham and McCormack 2004). On their part, artists have always been fond of art materials;[11] only recently, however, have art historians like Stafford (2007) linked the cognitive effects of images with biological mechanisms, indicating that mind and matter are closely linked on a personal, and, it should be added, on a collective level.

Material culture studies do explore matter in interdisciplinary manner (Fletcher 2010); paradoxically, though, certain aspects like the spirituality of matter are still largely circumvented, either because matter is seen as the crude opposite of spirit (Yourcenar 1985), or as out-of-bounds. This omission, which contrasts sharply with the ancient Greek heritage of matter as building stone of the cosmos, has major consequences, because it leaves a part of the urban experience beyond scrutiny.

Central Athens, for one, is an oneiric city vacillating between a charmed and a real order of things partly due to the veins, shades, forms, 'landscapes' and quivering surfaces of marble. The potential, dreams *and* nightmares of Athens are visibly articulated through marble and its mutilation, as I hope to further show later. It is appropriate that what was once theoretically unthinkable or liminal should come under the radar of analysis. Exactly this is the move I am making here: first, I explore the circumstances and impacts of public sacrilege against archetypal icons. Then, I study a number of dimensions inherent in the presence of elaborate marble artifacts which shape the urban space, art, and architecture of Athens, be them conscious – aesthetic, psychoanalytic, symbolic and metaphysical – or unconscious.

Recent uprisings in Athens have caused sustained injuries against marble architectural and sculptural monuments (Figure 2.3) through smashing, tugging, and even darkening whole façades of historical neoclassical buildings like the National Technical University of Athens.[12] Theorists like Gustave Le Bon, Elias Canetti, René Girard, and Louis Réau have shed light on the behaviour of crowds and on vandalism during periods of crisis, offering expedient insights for the analysis of impulsiveness, irritability, absence of critical judgement and emotional exaggeration. Girard, in his *Violence and the Sacred*, re-evaluates the concept of collective murder in Sigmund Freud's *Totem und Taboo* (1913), seeing ritual violence, sacrifice, and human scapegoating as the epitomes of the quasi-religious practices

Figure 2.3 The 'treatment' of the exquisite neoclassical Academy of Athens and of a later dismembered marble statue on the pedestrian zone of Tositsa Street next to the National Technical University of Athens.

© Photos and source: Argyro Loukaki.

necessary for the restoration of order in society. In the Athenian case, however, first, the marble-smashing collectivity is disputable. Second, destructive outbursts acquire the character of narcissistic delinquency since their time horizon is always open and 'spontaneous.' Third, one of the major movements in European cultural history, heritage restoration and protection which has flourished in the course of the 20th century, stirred responsiveness and initiated a lengthy as well as outstanding relevant state action in Greece (Loukaki 2008/2016) has not affected destroyers in the least. Fourth, victimages here are of two kinds, non-human and human. Alongside the scapegoating of some people, neoclassical monumental space has become a symbolic and real sacrificial scapegoat among groups pursuing the murder of the primordial city, parental or hosting.

Freud, in *Dostoyevsky and Parricide* (1928), says that 'before the problem of the creative artist analysis must, alas, lay down its arms'; however, the Athens center is regularly hit by destructive, not creative intention (for destructive creation in Athens, see next chapters). Today's tyrannical attacks on urban beauty and sublimity are actually attacks on the mythical, poetic, and profoundly political relationship of Athenians with the urban space. Animosity toward the neoclassical marble probably translates as a symbolic challenge to the Greek state with which it is identified. The loathing of specific groups against images of national collectivity abolishes degrees of freedom, nuances of urban excellence and imagined spatialities – Henri Lefebvre analyses persuasively their importance.

A lot of these actions are launched from and close to Exarcheia. A central neighbourhood, Exarcheia is home to many anarchists who initially shaped the anti-authority character of the neighborhood over the years, especially after the last dictatorship; a fact appreciated by many locals. This movement was reinforced in 2008, following the death of 15-year-old Alexis Grigoropoulos during a protest, caused by a policeman. The riots that rocked and burned Athens for weeks were a turning point. The situation escalated in the following years with the outrage caused by the economic crisis, resulting in a string of attacks.[13] In May 2010, during an anti-austerity protest, a Molotov cocktail bombing of a central bank set fire on a favourite neoclassical ensemble, resulting in the death of three employees, among whom a pregnant woman. Following this incident and the arson that destroyed a cultural icon like the movie theatre 'Attikon,' a previously glowing boulevard like Stadiou Street fell to partial dilapidation. Riots have caused further damage to the city's flesh, especially to marble façades, pavements, statues, benches, fountains, parapets, and stairs. With the neoclassical Athenian Trilogy looming large (Bastea 1999), monumental architecture and sculpture between the central Syntagma and Omonoia squares, where marble beauty is denser, have been constantly hammered through acts that classical Athens would punish as sacrilege and other modern cities would not tolerate. The severing of the urban body by a culture of constant rowdiness ultimately turns against the exhausted local society whose morale is hence broken even more. The heart of the Greek capital wrestles hard to respond to its unifying and symbolic role, weakened from the decentralization of public services and functions plus the 'decongestion' of the 1980s (Aravantinos 2002), the steep immigrant and refugee flux, as well as the visible signs of resident deprivation.

This is not the first time that design and material aspects of the urban form have concerned modern Athenians, including response to natural topography. Neoclassicism was locally disputed early on (Bastea 1999) when the small Athenian society felt ambivalent between, on the one hand, a strong desire for modernity and progress, which was spatially and visually portrayed by the closeness to classical paradigms, neoclassical street alignments, and the transformation of the early-19th-century small settlement to the capital city of Greece. On the other, the initiation of this sociospatial transformation not by local powers, but the environment of King Otto, son of the King Ludwig I of Bavaria. The latter, in fact, decided on the fortune of Athens as admirer of its history. Ambivalence was manifested as adherence to the former picturesque, irregular spatial code. Ironically, the city has followed these two codes, a linear and an irregular, since antiquity (see Chapter 1): Geometric regularity but with exquisite refinements and adjustments in its superior – in every sense – parts, which made the architectonic flesh live and breathing; organic urban fabric elsewhere.

The first city planning (1833), elaborated by architects Kleanthis and Schaubert, which formed the spatial basis for the transformation of Athens into capital of the new state, adjusted by Klenze, focused on the Acropolis Rock as visual and symbolic core (Figure I.1). Temples and buildings in its proximity, such as Thesseion (Hephaesteion) and Hadrian's Library were imitated by 'high' neoclassical complexes like the Athenian Trilogy.[14] Neoclassical diffusion through popular and picturesque applications lasted almost a century. The created warm, colorful, small-scale urban routine is still a stage for metropolitan theatricality which supports Athenians come what may. Postwar, the quarrying of Penteli and other peaks for building materials alarmed creators like Dimitris Pikionis because it threatened 'the divine line of (their) pediments' (Pikionis 1985, 131).

The history of European cities indicates that devastating urban crises have been transcended in surprising ways. Harvey (2003) showed how Paris, following the urban revolts that were put down horrendously, was subjected to urban restructuring, renovation, and innovation on all fronts through Baron Haussmann's vast planning, turning into the capital of 19th-century Europe. Such revolutionary urban changes tend to follow on the footstep of social revolts according to architectural historian Siegfried Giedion (1982). Europe at large was looking for new architectural paradigms after the revolutions of 1848. In France, the wide demand for architecture matching the dynamism of the times was expressed through the expediently great progress in metal structures as a result of the Industrial Revolution. Three remarks are necessary here: First, this European condition is not met this time around – far from it in fact. Second, immanent menaces do test cities but also push them to exceed themselves, if Harvey but also Herzog and de Meuron are right. What remains steady throughout the years are channels of adulation of this chthonic material, formal – especially the passion of specialists-, but also informal. Both thwart the present acts of violence in silent but persistent ways (see later in this chapter). Third, the absence of technological advancements of the French type is supplanted in Athens by the extraordinary burgeoning of the arts, including theatrical performances, Greek cineastes, exquisite new museums, new cultural centers, opera houses and composition, festivals, and new varieties of *arte povera*.

The Acropolis-Mount Pentelikon umbilical cord

The Acropolis and Mount Pentelikon were connected in classical times through the rugged road of marble, an umbilical cord that workers, architects, and artists ascended regularly. The road, a continual downhill from Penteli, followed the natural topography and was used to transport marble blocks from the quarry to the Acropolis.[15]

A passionate and complex rhizomatic relation arises in the Penteli-Acropolis *contrapunto*. The first pole (Figure 2.4), a beautiful mountain with some paradoxical features, transcendental landscapes, and ancient half-elaborated sculptures left *in situ*,[16] rich in vegetation and running waters, attracted modern Greek writers and naturalists who visited it regularly in Ruskin's adulatory manner, dedicating verses and literary accounts.[17] There are spiritual undercurrents in this mountain

Figure 2.4 The Athenian mountainscape from the top of Lycabettus Hill. Penteli is the pediment-like, faint mountain in the right side of the picture background.

© Photo and source: Argyro Loukaki.

of marble, once crowned by a statue of Athena, as witnessed the ancient geographer Pausanias.[18] Its bond with the Acropolis refutes the usual theoretical polarity between nature and culture; instead, it confirms the continuity of matter and spirit propelled by the Greek antiquity (see later in this chapter).

The Pentelikon cave (Figure 2.2), whence the marble for the construction of the Parthenon was extracted, one of the ancient quarries,[19] is a dense archetypal locus. A possible inspiration for the Allegory of the Cave in Plato's *Republic*, I would suggest, this cave represents labyrinthine unconscious, the workings of the conscious, initiation, philosophy, death and resurrection, gateways among worlds, inner vision, gnoseology, but also political constitution through emancipatory freedom. It has remained a place of timeless worship, ancient and Christian, revealing the dense manner in which cultural spacetimes mingle or emerge suddenly in Athens.[20] Tragedy, a vehicle of ecstasy and a school of democracy for

Figure 2.4 (Continued)

Athenians (Castoriadis 2007) was invented by poet Thespis of the Pentelic demos of Ikaria, present-day Dionysos.

The other pole, the Acropolis, was pinnacle of reason – just consider the mathematical calculations in its architecture, organization of the construction sites, plus the know-how of marble elaboration (see Korres 1992). The renowned 'breathing' of the Parthenon as a human body, attributed to optical illusions such as the designed bulging of columns (*entasis*) and the slight curvatures of architectural members (Coulton 1977; Korres 1992, 1999), was structurally feasible because the Pentelic marble can take extremely precise refinements. But the Acropolis was also seat of reverent myth; the Athenian unconscious nestled in the caves of its slopes (see Dodds 1951).

It accrues that the wider area of Athens is interweaved with venerable, solid relations, material and symbolic. This extraordinary richness is now being threatened and eroded.

Life in marble and its acolytes

Marble, a limestone metamorphosed into hard crystalline rock, keeps the general physical and symbolic properties of stone. Patterned texture, due to heterogeneity,[21] gives it a vast pictorial potential and a quasi-metaphysical appeal as both external presence and inward mass. Once revealed for the first time after being cut open, this 'inner' art records through quasi-scriptural, dotted or geometric forms millions of years of violent tectonic events that created it: fusions, tremendous temperatures and pressures, transformations, mixing of oxides, ruptures. Marble stones like *pietre paesine*, resembling Chinese landscape paintings or ruined cityscapes (Figure 2.5), evoke a primal quality, a deep aesthetic and cognitive world of elementary morphemes, objects and lifeforms, stirring the imagination

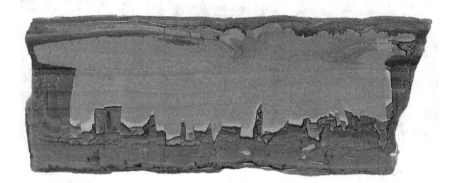

Figure 2.5 A pietra paesina.

© Photo: Wikipedia Commons. Source: https://it.wikipedia.org/wiki/Pietra_paesina#/media/File:Pietra_Paesina_di_Firenze_260x95_mm.jpg

of artists, poets and authors old and new, including Callois, Yourcenar, and Neruda.[22] Architects like Piranesi, aesthetes like Ruskin, philosophers like Leibniz and authors like Goethe were fascinated by mountains and rocks. Why? Possibly because, as excavation emerges human nature, so quarrying emerges the art of nature, eerily announcing, and even containing in minuscule form, all subsequent human art and architecture (see Caillois 1985). Such intuitions are corroborated by art historians and architects alike. Portoghesi explores the natural origin of building archetypes and considers architecture as the magical transcription of the first nature into the urban nature. He draws comparisons over the centuries revealing the unison between the human mind and the secret order of natural forms.

Artists respond ecstatically to this eternally pre-announced articulation of matter and spirit, commanding matter to release the unconscious. They are inspired by quarries, like Cézanne; treat marble like a living thing;[23] guess coherence and sense inner cracks, adapting their artistic visions accordingly. The symbolic load of marble is confirmed by Freud's psychoanalytic study of the statue of Moses in Rome and by his charged relationship with antiquity as paternal power (Loukaki 2008/2016, 2019).

Stafford (2007) has further explored this secret order.[24] Her understanding of the self-shaping tenaciousness of the natural world under tremendous pressure is substantiated in cut stones. Elegantly patterned self-assemblies of every imaginable shape and colour turn entropic dispersion into coherent syntax translated by human consciousness as a meaningful geometry of transient afterimages which announce a universal formal order. Various disciplines have coalesced over the past 150 years, corroborating the hypothesis of a higher coherence (Watson 2016).

Matter has been vested with spirituality at least since the Pre-Socratics initiated a great tradition exploring its metaphysics. Plato and more explicitly Aristotle were concerned with form and matter.[25] Hermetic books urged to 'listen to the great voice of things.' God Mithra is depicted emerging from a rock. Jesus used a stone metaphor in Thomas' Gospel: 'Split a piece of wood – I am there. Lift the stone, and you will find me there.' Romantics believed that visual formulas such as those of cut stones reveal an ingrained mentality.

Building materials, the elementary substance of urban life, carry sequences of nuances tacitly. The familiar forms they bear quiver gently on all kinds of urban surfaces, establishing relations of sentient immediacy to surrounding spaces and realities, linking us to respective natural forms, their structure, their particular radiance. Materials in conjunction with colour accomplish formal clarifications, redefinitions and individualizations (Chimonas in Antonakakis 2013, 27–29). Marble skin and inner life promulgate intersubjective though unvoiced experience. Sensorial, in conjunction with historical experiences, explain the deep fascination of this Athenian stone, while contributing toward social coherence, the public aesthetic, and collective meaning.

Noble both in full architectural or sculptural forms and in fragmentation, which paradoxically highlights workmanship and material quality, marble in Greek temples gives geometrically sharp, exquisite ruins. The structural system of marble temples geometricizes the sky when seen against it, contrasting human measure, drive, and creativity to the cosmic order. These open dialogues with light launch aesthetic categories. In comparison, brick-and-concrete Roman ruins, ominous and gaping, activate the uncanny as aesthetic category. Difference between geometric clarity and porous mystery differentiates sharply the archaeological landscapes of Athens and Rome, the two archetypal classical cities (Loukaki 2019).

Part of the debt of Modern to classical Greece hinted at previously is repaid through innovative restoration programs, the construction of new museums like the Acropolis Museum and care for the immediate environment of the Acropolis, including the exquisite archaeological landscape of Pikionis. The architect elaborated, weathered, and layered marble in ways evoking descent from the deep past or synergy with modern materials like concrete in suitably chosen sites of his landscape.[26]

Sculpture draws its objects to light from the depths of matter, mediating between the artist and the divine, as illustrated in the myth of Galatea and Pygmalion. With marble statues like the winged Nike of Samothrace or Aphrodite of Melos (Venus de Milo), the fusion of an archetypal human and a heavenly world, the sublime epiphany manifested through extraordinary human talent, representing a certain collectivity in time-space, coupled with eroticism, flabbergasts spectators with unprecedented emotion.[27] Great sculptors like Phidias (Figure 2.6) and

Figure 2.6 The statues of the eastern pediment of the Parthenon, sculpted by Phidias, in the British Museum.

© Photo: Wikipedia Commons. Source: Elaborated from https://en.wikipedia.org/wiki/Parthenon#/media/File:Pediments_of_the_Parthenon-British_Museum.jpg

Michelangelo explore the morphoplastic resistance of marble while struggling to release its always emerging spirit, linking us with possible worlds,[28] and reflecting the creative moulding of mental effort. This, quite possibly, explains why Hegel believed it is hard to understand ancient philosophy and history without studying sculpture.

The use of white marble to represent the human body is age-long. Nude, wet or wind-blown, the body was uniquely rendered with this material. Statues, ideal Doppelgängers of real persons, pulsating and almost alive, appear to be moulded from the inside out as if shaped by a life-giving power. Enveloped by the aura of authenticity and similarity, they sharpen analogous thinking beyond the physical body and re-enact myth constantly.

Painting as a representation of the spatial pair Pentelikon-Acropolis has dealt with the pictorial qualities of both, especially after the founding of the Greek state in 1832. Painters pursued the deeper qualities of marble empowered by the classical past and the ancient bonds between painting and poetry.[29] The Sacred Rock attracted, as would be expected, many more artists than Penteli, both Greeks and foreigners. There are, however, differences in approaches: foreigners inspect, study, learn lessons, and still perceive the East in Athens, or restore a brilliant past for the West in Greece, especially during the 19th century. This is quite Klenze's move in his painting *Reconstruction of the Acropolis*. Greeks like Vassiliou here (Chapter 13) maintain with the Acropolis marbles an existential relationship: they worship them, draw inspiration, resist them in awe, explore their role in modern Greek community, their greatness, or their future position in a transforming urban fabric. Besides, Greek painters comment on previous works by foreigners; perceive the Acropolis as absolute heart of the city despite time and vicissitudes; pose, and answer, the question 'to whom this place belongs.'

The popular worship of marble in Athens involves the performance of various urban rituals whether in holiday or daily routine, next to, opposite, or even on marble fragments of architectural members (Figure 2.7). During days of limpid brightness, when archaeological sites turn into urban ecosystems of exquisite beauty, ancient fragments, hosting the Athenian feline wildlife, enrich the local visual unconscious with images of preciously anarchic minglings as part of the Athenian identity. Local Byzantine monuments had already used ancient recycled marbles, sometimes, conjoined with Byzantine elaborate members as it happens in the central 'Little Metropolis' (Panagia Gorgoepekoos). Present adhesion to marble includes burial use, following the ancient and neoclassical template;[30] the marble table-tops in the cafés where convened the creative generations of the 1930s and 1960s; pavements of the historic centre; modern façades. The sweeping presence of this material distinguishes central Athens from other European cities.

It is now hopefully becoming clearer why marble, dominant in the monumental art and architecture of Athens, condenses sequences of aesthetics, local topographies, properties, functions, symbolisms, and emotions. Why it motivates infinite folds of meanings: psychic vibrations, sensory memories,

Figure 2.7 Flying kites on Clean Monday on Pnyx against the Acropolis, Lycabettus, and Hymettus.

© Photo and source: Argyro Loukaki.

mythologies of urban constitution, daily encounters with archetypes, close associations with the past.

Attitudes toward the past, from the most glorifying to the most iconoclastic – and there are all kinds in Athens- indicate that it is impossible to escape measuring up with it. The material background of a city like Athens does affect its space, art, social life, urban memory, identity, but also national and local dignity to an extent that has not yet been effectively appreciated. Attacks against it, especially when persisting for so long, impact negatively the collective body, impoverishing or undermining its experience on all these fronts. The morale and cohesion of the Athenian society are injured exactly when most needed; this can hardly be emancipatory political struggle worthy of its name. The certain outcome of the urban disfigurement combined with the preceded abolition of important institutions such as the Unification of Archaeological Sites, an impact of the crisis, is that it has facilitated the transformation of Athens into prey of global capital due to the fall of land value.[31]

When no cultural rules can hold anymore, culture moves toward dissolution, warns Guy Debord (1994). Athens has but restoration of the creative atmosphere and the urban wholeness on all fronts of the urban life does appear imperative now. This in Athens depends on the recuperation of elementals like urban beauty and sublimity. The kind of architectural certainty that so fascinated modern painter Yannis Tsarouchis as a secure, familiar and peaceful backdrop (Loukaki 2009), including modernism, local urban gardens with their poets' statues, official neoclassical architecture, as well as small-scale neoclassical ensembles injecting warm humanity within bigger neighbourhoods, can contribute to this restoration. Such an urban routine would be inspiring and heartening if we judge from the cultural rules applying in early 1960s Athens, a city that freed possibilities, respected its materiality and cultivated creative fever. In a climate fulfilling a number of some more positive preconditions, a major take-off in the arts took place.

Says Dante:

> Midway upon the journey of our life
> I found myself within a forest dark,
> for the straightforward way had been lost.

And Elytis answers:

> What is good? What is bad?
> A point
> And on it you balance and exist
> And beyond it agitation and darkness

Both poets induce the idea that it is necessary to find a measure and balance between the need for resistance to the economic undoing of a country (see Introduction) and the demolition of its capital city's culture, both material and immaterial. It appears absolutely vital for a nation haunted by traumatic loss to nurture the collective consciousness, namely to agree upon the current values of the *koinon* (common), the beautiful, the sublime and the main, so important to the antiquity but also to a founding figure of Modern Greece, national poet Dionysios Solomos. It is also vital to agree upon the decisive role of the demos, of theoreticians, activists and artists, Greeks and foreigners. A locus of consensus, creative and purifying, is essential in this midway crossroads.

Notes

1 In Pericles' Funeral Oration from Thucydides' *History of the Peloponnesian War*.
2 They are studied consistently by Greek and foreign specialists. Archaic votive sculptures plus the pedimental sculptures of the Acropolis Old Temple were of Parian marble. While island marbles like the Parian were imported in the 6th century, local quarries were also developed on Pentelikon and Hymettos. Early in the 5th-century marble started being used for whole buildings, mainly temples (Wycherley 1974). The

Parian *lithos* was never entirely abandoned as raw material for sculpture in classical times (Palagia 2000). See also Korres (2001).

3 This, plus the quality of elaboration, differentiates Athenian neoclassicism from the neoclassical mansions of Hermoupolis, capital of the Cycladic island of Syros, which imported marble from the neighbouring islands of Tenos and Paros.

4 See Rossi (1990); Halbwachs (1992); Boyer (1994).

5 http://hybridpedagogy.org/from-under-volcano/

6 Even though classical Athens was predominantly a city of mud brick (Wycherley 1974) on stone foundations.

7 For Ginsberg, see http://radiobookspotting.blogspot.gr/2015/06/allen-ginsberg-3-1926.html. For Merrill, see Hammer (2015); also Loukaki (2008/2016).

8 Song by Mikis Theodorakis based on a poem by Yannis Ritsos, translated by the author.

9 Song by Mikis Theodorakis based on a poem by George Seferis, translated by the author.

10 See, indicatively, Harvey's *Justice, Nature and the Geography of Difference* and *Limits to Capital*, Smith's *Uneven Development*, and Swyngedouw's *Liquid Power: Contested Hydro-Modernities in Twentieth-Century Spain*.

11 See the physical manner the art historian and painter James Elkins (2000) relates to materials in paintings.

12 There is the precedent of the events of November 17, 1991, when the building of the National Technical University Rectorship was set on fire during the exchange of Molotov cocktails and tear gas between protesters and special police officers. The building and valuable artworks were destroyed.

13 Wolcke (2019).

14 Analysis in Kienast (1995).

15 This road has been explored and fully documented by the chief Acropolis restoration architect, Manolis Korres (1994).

16 For details, see Oresivios (2012).

17 Ruskin (2010) was fascinated by the Alps. Greek authors include Dimitris Kampouroglou, Kostas Krystallis, Zaharias Papantoniou, and Elias Venezis, among others.

18 The two possible places of the Athena statue are either in the area of the ancient quarries, next to the cave, or close to the mountain top.

19 The emergence of the cave in classical times following quarrying and the exceptional organization of the ancient quarries in Mount Pentelikon, possibly based on the precedent of the Paros quarries, are discussed in detail by Korres (1994).

20 There are a large number of ancient sanctuaries and monasteries on Pentelikon, both Byzantine and modern, see Oresivios (2012).

21 Absolute, 'dead' whiteness is fully realized in stucco imitations (Wycherley 1974).

22 Orange stains . . . of oxide
green veins on the calcareous peace
that foam mints with its wrenches
or dawn with its rose,
thus are these stones:
nobody knows whether they emerged from the sea or to the sea return,
something caught them while they were alive,
they expired in immobility
and built a dead city.
A city without cries, without kitchens,
a solemn enclosure . . . of purity,
pure forms fallen
in a disorder without resurrection,

in a multitude that lost its gaze
in a grey monastery
condemned to the naked truth of its gods.
Source: www.lanuovamusiva.com/en/tuscan-stones-collection/
23 As shown by anthropological research (Leitch 1996).
24 See Stafford (2007, 187) and sporadically.
25 www.newworldencyclopedia.org/entry/Form_and_Matter (Accessed 20.07.2019).
26 Analysis in Loukaki (1997, 2014/2016).
27 . . . I stopped in front of The Victory of Samothrace. The Winged Victory. I stayed
   there, standing, flabbergasted. That figure was soaring, like an airplane in a silent
   dream . . . A civilization was in front of my eyes taking off and still remaining on
   Earth, a civilization. . .(with) its archetypal manifestation. And then I turned and saw
   the Venus de Milo. The statue of white marble was standing there, rising or bending
   according to one's place, seemingly larger than life . . . The love that possessed the
   Greek sculptor in bringing this goddess to form, to shape, to life, moved over to me
   (Adnan 2011, 9).
28 See Harvey (2000).
29 Dialogues between painting and poetry were established in the Archaic era (8th century BC to 480 BC). Simonides of Ceos called painting silent poetry and poetry painting that speaks.
30 Of the pioneers from the Cycladic island of Tenos.
31 www.in.gr/2019/11/12/economy/oikonomikes-eidiseis/pou-agorazoun-akinita-oi-kse
   noi-ependytes-ti-deixnoun-ta-stoixeia-tis-golden-visa/ (Accessed 12.11.2019).

Bibliography

Adnan, E. 2011. *The Cost for Love We Are Not Willing to Pay: 100 Notes, 100 Thoughts.*
   Documenta Series 006. Berlin: Hatje Cantz.
Alesch, J. 2007. *Marguerite Yourcenar-The Other/Reader.* Birmingham, AL: Summa
   Publications.
Antonakakis, D. 2013. *Dyo Dialexeis gia ton Pikioni-Two Lectures on Pikionis.* Athens:
   Domes.
Aravantinos, A. 2002. 'Dynamikes and Schediasmos ton Kentron stin Pole ton Epomenon
   Dekaetion – Pros Sygentrotika e Apokentrotika Schemata? (Dynamics and Planning of
   Urban Centers in the City of the Next Decades – Towards Centralizing or Decentralizing
   Schemes?).' In *Aeichoros* 1(1), 6–29.
Bachelard, G. 1994. *The Poetics of Space.* Boston: Beacon Press.
Bastea, H. 1999. *The Creation of Modern Athens: Planning the Myth.* Cambridge: Cam-
   bridge University Press.
Benjamin, W. and Lacis, A. 1978. 'Naples.' In Benjamin, W. ed., *Reflections.* New York:
   Harcourt Brace Jovanovich, 163–173. Available at: www.columbia.edu/itc/architecture/
   ockman/pdfs/session_8/benjamin.pdf (Accessed 28.05.2019).
Boyer, M.C. 1994. *The City of Collective Memory.* Cambridge, MA: The MIT Press.
Callois, R. 1985. *The Writing of Stones.* Charlottesville: University Press of Virginia.
Castoriadis, C. 2007. *The Greek Particularity-From Homer to Heraclitus.* Athens: Kritike.
Coulton, J.J. 1977. *Ancient Greek Architects at Work: Problems in Structure and Design.*
   New York: Cornell University Press.
Dagi, D. 2018. 'European Union's Refugee Crisis. From Supra-Nationalism to National-
   ism?' In *Journal of Liberty and National Affairs* 3(3), 9–19.

Damaskos, D. and Plantzos, D. eds. 2008. *A Singular Antiquity: Archaeology and Hellenic Identity*. Athens: Benaki Museum.

Debord, G. 1994. *The Society of the Spectacle*. Cambridge, MA: Zone Books.

deLillo, D. 1992. *The Names*. New York: Alfred A. Knopf.

Dodds, E.R. 1951. *The Greeks and the Irrational*. Berkeley and Los Angeles: University of California Press.

Elkins, J. 2000. *What Painting Is*. London: Routledge.

Fletcher, R. 2010. 'Urban Materialities: Meaning, Magnitude, Friction, and Outcomes.' In Hicks, D. and Beaudry, M.C. eds., *The Oxford Handbook of Material Culture Studies*. Oxford: Oxford University Press, 459–483.

Freud, S. 1913. *Totem und Taboo*. Boston: Beacon Press.

Freud, S. 1928. 'Dostoevsky and Parricide.' In *The Realist* 1(4), 18–33.

Giedion, S. 1982. *Space, Time and Architecture* (first edition 1941). Cambridge, MA: Harvard University Press.

Girard, R. 1978. 'Diacritics.' In *Special Issue on the Work of Rene Girard* 8(1), 31–54.

Girard, R. 1979. *Violence and the Sacred*. New York: W.W. Norton & Company.

Halbwachs, M. 1992. *On Collective Memory*. Chicago: University of Chicago Press.

Hameroff, S. and Penrose, R. 2014. 'Consciousness in the Universe.' In *Physics of Life Reviews* 11(1), 39–78.

Hammer, L. 2015. *James Merrill-His Life and Work*. New York: Alfred A. Knopf.

Harvey, D. 2000. *Spaces of Hope*. Edinburgh: Edinburgh University Press.

Harvey, D. 2003. *Paris, Capital of Modernity*. London: Routledge.

Harvey, D. 2007. *The Limits to Capital*. London: Verso.

Kienast, H. 1995. 'Athener Trilogie. Klassizistische Architektur und ihre Vorbilder in der Hauptstadt Griechenlands.' In *Antike Welt* 26(3), 161–176.

Korres, M. 1992. *The Construction of Ancient Columns*. Unpublished doctoral thesis, School of Architecture. Athens: National Technical University of Athens.

Korres, M. 1994. *From Pentelikon to the Parthenon*. Athens: Melissa.

Korres, M. 1999. 'Refinements of Refinements, in: Appearance and Essence. Refinements of Classical Architecture: Curvature.' In Haselberger, L. ed., *Appearance and Essence*. Philadelphia: The University of Pennsylvania.

Korres, M. 2001. *From Pentelicon to the Parthenon: The Ancient Quarries and the Story of a Half-Worked Column Capital of the First Marble Parthenon*. Athens: Melissa.

Latham, A. and McCormack, D.P. 2004. 'Moving Cities-Rethinking the Materialities of Urban Geographies.' In *Progress in Human Geography* 28(6), 701–724.

Le Bon, G. 2009. *Psychology of Crowds*. Southampton: Sparkling Books Limited.

Lefebvre, H. 1992. *The Production of Space*. Oxford: Blackwell.

Leitch, A. 1996. 'The Life of Marble: The Experience and Meaning of Work in the Marble Quarries of Carrara.' In *The Australian Journal of Anthropology* 713, 235–257.

Loukaki, A. 1997. Whose *Genius Loci*? Contrasting Interpretations of the Sacred Rock of the Athenian Acropolis. In *Annals of the Association of American Geographers* 87(2), 306–329.

Loukaki, A. 2008/2016. *Living Ruins, Value Conflicts*. London: Routledge.

Loukaki, A. 2009. 'Peiraias: Idanikes Ekdoches tou Astikou Topiou sti Zografike tou Gianni Tsarouche' (Piraeus: Ideal Versions of the Urban Landscape in the Painting of Yannis Tsarhouchis).' In Kalafatis, T. ed., *Peiraias: Historia kai Politismos (Piraeus: History and Culture), Proceedings of Conference*. Piraeus: University of Piraeus, 214–261.

Loukaki, A. 2011. 'Nesiotekoteta kai Anaptyxe: Astike Stromatografia kai Apoklinouses Ermeneies tou Athenaikou Protypou stis Poleis tes Kretes (Islandness and Development:

Urban Stratigraphy and Divergent Interpretations of the Athenian Paradigm in the Cities of Crete).' In Demathas, Z. ed., *Tomos Timetikos gia ton Kathegete Pavlo Loukake (Volume in Honour of Professor Pavlos Loukakis)*. Athens: Dardanos, 604–643.

Loukaki, A. 2014/2016. *The Geographical Unconscious*. London: Routledge.

Loukaki, A. 2019. 'Central Archaeological Domains of Athens and Rome: Physical, Metaphysical and Moral Aspects of their Respective Management.' In Moraitis, K. and Rassia, S. eds., *Urban Ethics Under Conditions of Crisis: Politics, Architecture, Landscaping and Multidisciplinary Engineering*. Singapore: World Scientific.

Oresivios. 2012. *Pentele (Penteli)*. Available at: https://iranon.gr/PO/penteli.pdf (Accessed 22.05.2019).

Palagia, O. 2000. 'Parian Marble and the Athenians.' In Katsarou, S., Katsonopoulou, D. and Schilardi, D.U. eds., *Paria Lithos (Parian Stone)*. Archaiologikes kai istorikes meletes Parou kai Kykladon.

Pericles 'Funeral Oration' from Thucydides.' In *History of the Peloponnesian War*. Available at: https://sourcebooks.fordham.edu/ancient/pericles-funeralspeech.asp (Accessed 24.02.2020).

Pikionis, D. 1985. *Texts*. Athens: Educational Foundation of the National Bank of Greece.

Réau, L. 1994. *Histoire du Vandalisme*. Paris: Robert Laffont.

Rossi, A. 1990. *The Architecture of the City*. Cambridge, MA: The MIT Press.

Ruskin, J. 2010. *Praeterita*. Oxford: Oxford University Press.

Smith, N. 2008. *Uneven Development*, 3rd ed. Athens: University of Georgia Press.

Stafford, B.M. 2007. *Eco Objects-The Cognitive Work of Images*. Chicago: The University of Chicago Press.

Swyngedouw, E. 2015. *Liquid Power: Contested Hydro-Modernities in 20th-Century Spain*. Cambridge, MA: The MIT Press.

Washington, H.S. 1898. 'The Identification of the Marbles Used in Greek Sculpture.' In *American Journal of Archaeology* 2, 1–18.

Watson, P. 2016. *Convergence-The Idea at the Heart of Science*. New York: Simon & Schuster.

Wolcke, A. 2019. 'The End of Anarchists in Exarcheia?' In *European Interest*. Available at: www.europeaninterest.eu/article/end-anarchists-exarcheia/ (Accessed 08.02.2019).

Wycherley, R.E. 1974. 'The Stones of Athens.' In *Greece & Rome* 21(1), 54–67.

Xenophon. *Poroi*. Available at: https://stuff.mit.edu/afs/athena/course/21/21h.401/www/local/xenophon_poroi.html (Accessed 23.02.2019).

Yourcenar, M. 1985. 'Introduction.' In Callois, R. ed., *The Writing of Stones*. Charlottesville: The University Press of Virginia, xi–xix.

Part 2

The artistic sublime of the Byzantine Cosmopolis

Between imitation of classicism
and the Christian Dogma

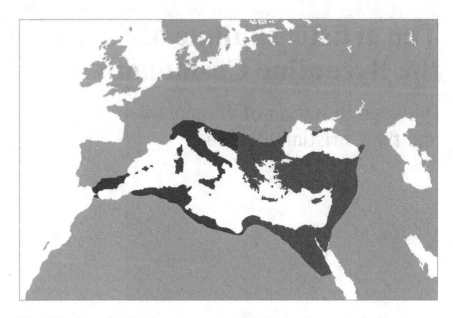

Map 2 The Byzantine Empire at its greatest extent in the 6th century. Territories were reconquered during the reign of Justinian I.

Source: © Wikipedia Commons. Source: https://commons.wikimedia.org/wiki/Category:Maps_of_the_Byzantine_Empire_in_565#/media/File:ByzantineEmpireGE.PNG

Diagram 1 Constantinople in Byzantine times.

© Diagram: Wikipedia Commons. Source: http://en.wikipedia.org/wiki/User:DeliDumrul

3   On real and imaginary cities

Textual and visual representation of cities and the perception of urban space in the Byzantine world

Vicky Foskolou

In the early early 14th century, Theodore Metochites, one of the most important political personages and a prolific writer of the late Byzantine era, composed an *encomium* for Constantinople, the city of his birth. In a focal point of his text, Metochites describes Hagia Sophia and expresses the belief that the church of God's Wisdom is a creation before which the seven wonders of the ancient world fade into insignificance. In his high sophisticated style he wonders 'which of the ancient wonders of the world can be compared with such a powerful edifice without occasioning laughter or people calling us simpletons? What walls of Babylon . . . Egyptian pyramids, mausoleums in Karia, or hanging gardens. . .' (Politis 2013, 343–353).

Comparing the contemporary monuments of a city with famous buildings from antiquity is one of the topoi, that is, the literary formulas, frequently found in such texts (Rhoby 2012). This is no coincidence but is due to a large extent to the roots of the encomia and *ekphraseis* on cities in the art of rhetoric, to which they also owe their more or less standardized structure (Pernot 1993, 181–216).[1] Panegyrics and descriptions of cities were based on the rules found in rhetorical handbooks, the best known example being the late 3rd century AD treatise *How to Praise a City* by Menander the Rhetor. According to these instructions, a city should be praised for its position and climate, its landscape and the fertility of the soil, the origin and deeds of its inhabitants, and finally for the beauty of its monuments and temples (Saradi 1995). Common literary motifs and the similar structure of the rhetorical descriptions of cities also owe much to the fact that earlier famous texts with analogous content were used as models and became a source of inspiration for later generations of authors.[2]

Encomia and *ekphraseis* of cities were written either as independent texts, usually orations given on some special occasion, or as descriptions of urban landscapes embedded in various types of literary works. They were particularly widely distributed in late antiquity, that is, up to the 6th century AD, and experienced a revival in the late Byzantine period from the 13th century on (Hunger 1978; Saradi 2011).

There is nothing coincidental about the periods when such texts were most widespread. It is a historical conjuncture that reflects the place and importance of cities in these periods. More especially, in late antiquity the Roman Empire was perceived as a system of cities that were a byword for the civilization and traditions of the Graeco-Roman world. As Eleni Saradi has pointed out, two elements determined both the textual and pictorial depictions of the late antique Roman world: the sea and the cities (Saradi 2010). Thus, in Procopius's *Buildings* Justinian's empire is projected as a group of cities that are reconstructed, restored, and fortified (Saradi 2006, 71–79). Similarly, the depictions of cities we frequently come across on mosaic pavements in private houses and churches give a panorama of the urban centres of the empire, thus promoting not only the values of civic life but also the spread of political and ecclesiastical power in this period (Saradi 2006, 118–146, 2010).

Figure 3.1 Jerusalem. Detail from the floor mosaic of the church of St Stephen in Umm al-Rasas, Jordan. Early 8th century AD.

Photo: Dick Osseman.

Visual representations of cities were very widespread on mosaic pavements in churches of the Middle East. The most important examples are found in the basilicas of Gerasa and Madaba in Jordan and are dated to the 5th and 6th centuries, but they continue up to the early years of the Arab conquests, as in the case of the early 8th-century mosaic floor in the church of St Stephen in Umm al-Rasas, where the border of the composition is adorned with more than 20 images of named cities (Bowersock 2006) (see Figure 3.1). The mosaic miniatures of towns on the floors of Christian basilicas have been approached from various points of view and consequently a variety of queries have arisen. Are they conventional iconographical subjects with a purely decorative role (Saradi 2006, 2010), or could actual topographical details of the cities be hiding in these architectural vignettes that often surround larger pictorial compositions (Bowersock 2006)?[3] Are they referring to a biblical or more specifically Christian geography (Saradi 2006, 2010), or are they aiming to portray the inhabited world and landscape of their day according to a Roman world – and perhaps political – view (Bowersock 2006)? Wherever the answers lie, the plethora of similar pictorial examples reveals the crucial role played by cities in the society of the period. This view is borne out by the iconographic motif of the Tyche of the city, that is, the female figure that personifies a city and bears its symbols. The examples from this period continue the tradition of Hellenistic times in an unbroken chain up to the late 6th century (Gittings 2003; Poulsen 2014). Typical of the genre are the elegant, small silver gilded statuettes of the personifications of Antioch, Rome, Constantinople, and Alexandria from the Esquiline treasure, now in the British Museum, once sumptuous decorative fittings for furniture, which reveal the spread of this iconographic motif to the everyday lives of the urban upper classes.[4]

After a long break the encomiastic descriptions of cities reappear, as mentioned previously, in the late Byzantine period. This has been associated with the new political conditions created by the break-up of the Byzantine Empire after 1204 by the crusaders. In the politically fragmented world of the Byzantine territory new centres of power would emerge outside Constantinople and the contacts with the West would contribute to a more general economic development and social reinforcement of the provincial cities (Saradi 2008; Kiousopoulou 2013). So it hardly seems a coincidence that more than one encomiastic description was written at that time about Nicaea, the capital of the Byzantine Empire in exile, and for Trebizond, the center of power of the Great Komnenoi in the Pontos, emphasizing their magnificence and importance (Saradi 2011, 2012b; Voudouri 2016).

Nevertheless, the flourishing of literary descriptions is not paralleled by a similar effort in visualizing urban landscapes in late Byzantine art. Although Palaiologan art was characterized by an intense interest in defining pictorial space using buildings and elements of the natural world (Velmans 1964), what in the end is represented in painting are complicated architectural frameworks, surrounded by nature in bloom and bedecked with rumpled hangings and picturesque decorative details. Thus, they are more reminiscent of a theatrical backdrop (Vasilakeris 2014) than realistic landscapes and cities, even of a symbolic or imaginary nature, such as the ones that appear in contemporary Italian works.[5]

The motif of an enclosed fortress, from which densely packed rooves of houses with the dome of a church in their midst projected, which constituted the typical iconographical convention for depicting cities in Byzantine visual culture from the 8th century on, remained in use up to the end of the empire in the mid-15th century (Saradi 2010). The impressive fortified city in the scene of the Entry into Jerusalem in the Pantanassa Church in Mistra (ca. 1428–1444), one of the last great painted programmes of the Byzantine tradition, underlines the enduring presence of this iconographic subject in Byzantine art (see Figure 3.2).[6]

As mentioned previously, Byzantine writers and orators faithfully followed the compositional guidelines for the particular literary genre and drew on earlier rhetorical tradition, resulting in descriptions of cities in Byzantine texts having an almost standardized form and uniform construction. As Tasos Tanoulas has characteristically commented: 'Landscapes in Byzantine *ekphraseis* are constructed from the same turns of phrase, set out on the same grid, within which they occupy equivalent spaces and, as a result, look as if they have been "cast in a mould."'[7]

Figure 3.2 Entry into Jerusalem. Wall painting, ca. 1428. Church of the Pantanassa, Mistra.
© Photo: M. Kappas.

Indeed, the texts focused on the beauty and prosperity of the city concerned by presenting its geographical location, its abundant natural riches and ideal climate. The idyllic natural scenery that surrounds the city is then sketched in with words. Plants, flowers, trees, vineyards, running water, lakes and seascapes are described by the authors of encomiastic *ekphraseis* with lyrical imagery (Saradi 2012). After that reference will be made to its fortifications, since this was synonymous with strength, security, and the very size of a city. Walls and towers are glorified for their location, dimensions, the symmetry of their design, their firm build and splendid fabric, things that give the city's inhabitants security, pride and aesthetic pleasure and put fear into their enemies (Saradi 2011; Rhoby 2012, 88–91). The emphasis on the beauty and the divine protection provided by the fortifications reveals the aesthetic value and symbolisms that surround the walls in the eyes of the Byzantines (Maguire 1993, 94; Bakirtzis 2012). There is no doubt that the walls of Byzantine cities, with their votive inscriptions that extol the leaders who built and repaired them, the scattered spolia immured in their façades reminding people of the historical continuity of the place and the apotropaic symbols that adorn them, acquire hierotopical dimensions and separate 'the familiar from the hostile, the known from the unknown, the good from the bad' (Bakirtzis 2012, 157). From this point of view it is interesting to note that in visual representations of walled cities, despite their miniature and abbreviated form, a saintly figure in bust form often appears above the central gate, an iconographic detail that emphasizes the divine protection the walls gave the city and its inhabitants (see Figure 3.3).[8]

Geographical location, natural surroundings, and fortifications are consequently the building blocks of literary descriptions of cities. The short, mid-14th century *ekphrasis* on Trebizond from the pen of Andreas Libadenos confirms this deduction, containing all three of these in just two paragraphs:

> And the city has a wonderful location . . . Likewise, it gives the people there and even the animals everything they could need, and indeed in great abundance. It inspires fear in the people of the hinterland with its walls, which look to them as if they were made of brass. For locals, strangers and traders it provides plenty of precious goods. The air there is also good as are its waters, which flow from both sides . . . This means that all sorts of herbs grow there and it can boast thousands of flowers and is filled with an abundance of fruit; and the place is constantly growing more beautiful thanks to the changing flowers, while along the coastline it is crowned with flat rocks and sandy beaches
>
> (Lampsidis 1975, 61, ll. 12–29)[9]

As regards the urban space surrounded by walls, a focal point of the *ekphraseis* was the houses, emphasizing their size and the quality of construction, and pointing out how densely packed they were, all things that endowed the city with special grandeur. The writers paint an interesting image in words, describing these high-rise homes that 'dispute with the wind,' giving the impression that part of the

Figure 3.3 The Angel smiting the Assyrians before the city of Jerusalem. Wall painting,
Constantinople, Chora Monastery (1320–21).

Photo: Dick Osseman.

city is built in the sky.[10] The noble residences that have been preserved in good
condition in the late Byzantine city of Mistra give us a picture of these grand mansions mentioned in the texts (Kalopissi-Verti 2013).[11]

However, what the authors of these urban *ekphraseis* linger longest over is
the description of religious buildings to which a city usually owes its reputation
(Saradi 1995) and which is also the only detail to stand out in visual images of
towns.[12] Indeed the texts focus on the church buildings with such emphasis that
in some cases they become metonymous with the cities (Messis 2012). Haghia
Sophia and Constantinople is perhaps the most typical example of this interrela-
tionship, the quasi-identification of a city with its main church. Symbol and trophy

of the empire and the city, its beacon, heart and eye on the world, are just some of the literary metaphors for Haghia Sophia found in the encomia on the imperial capital (Fenster 1968; Rhoby 2012). Pictorial parallels of these metaphors can be found in images of Constantinople in illustrated manuscripts that represent the city as a fortress with only one domed church inside the walls, namely Hagia Sophia.[13]

It was not only the church buildings that were identified with a town, but also the saints to which they were dedicated, as the case of St Demetrios and Thessaloniki aptly demonstrates. It is scarcely necessary to point out that Demetrios and his basilica play a central part in literary descriptions of Thessaloniki (Konstantakopoulou 1996; Kaltsogianni et al. 2002). What is interesting, however, is how in these texts both the saint and the city are presented as living beings, virtually human, acting and interacting. Thus, as is characteristically mentioned in one of these *ekphraseis*: 'Thessaloniki had entrusted its safety to Demetrios and he in turn fought zealously on its behalf.'[14] The special relationship between the saint and his town is vividly illustrated in a series of scenes of his life in Serbian monuments that represent Demetrios's miraculous interventions to save the city. Indeed, in one of them he is shown majestically striding along the battlements of the walls of the city, repelling its enemy (Stojaković 1966; Bauer 2013).[15] A Cretan icon from the end of the 16th century, now in Corfu, that depicts Demetrios on horseback in front of a walled city, identified by an inscription as Thessaloniki, is also another visual proof of the strong bond between the young military saint and the second city of the Byzantine Empire (Ćurčić and Hadjitryphonos 2010, no 34) (see Figure 6.2).

Aspects of the urban landscape connected with the activities and economic life of its inhabitants are also mentioned in the rhetorical descriptions and praises of cities, in some cases creating vivid and colourful imagery. One such text states that 'it was easier to count the grains of sand on the beach than the [number of] people crossing the marketplace or involved in trade.'[16] Nevertheless, this aspect of cities is rarely shown in pictorial representations. The only exception, which is all the more important for that, is the late 13th-century representation of the procession of the icon of the Virgin Hodegetria in Constantinople, located in the Blacherna Monastery church in Arta, Epirus (Parani 2016). The composition also vividly depicts some rare details of the festive atmosphere prevailing as the icon is carried in procession and the merrymaking in the precincts of the church (see Figure 3.4). Men and women participate in the procession, which takes place in an open space in front of a building with several storeys, with female figures hanging out of the windows and over the balconies, while in the foreground we see pedlars displaying their wares: a woman selling vegetables, a man selling caviar and an old woman with small cups hung around her neck for selling wine to the assembled company. This scene allows us to catch a glimpse of the area of the marketplace, where the inhabitants of Byzantine cities could gather, not only to exchange produce or take part in a religious ceremony but also to drink, enjoy themselves and get together with their fellow citizens.[17]

Figure 3.4 The procession of the icon of the Virgin Hodegetria. Wall-painting, Blacherna
    Monastery church Arta, Epirus. Late 13th century.

Photo: M. Kappas.

Yet this scene is a one-off. So, with texts and visual representations giving only
idyllic images of cities, encircled by high walls, adorned by important buildings,
and set in fertile natural surroundings, the question arises: Could these 'cast-in-
a-mould' texts and painted vignettes offer us any information at all about urban
space in Byzantium?

The answer is that what they give us above all is the feeling or even the emotions evoked by the city in the writers and in their readers; in other words, how
authors and audiences perceived the notion of a city and its landscape in a specific
historical context (Tanoulas 2012).[18] Thus the literary images of cities with high
walls and densely packed houses reflected a sense of an enclosed, inward-looking
and thus safe urban space, while the same feeling is evoked in the viewer by
the Byzantine visual 'cityscapes' (Kiousopoulou 2013). Moreover, each author
focuses in a different way and with differing emphasis on the details that make
up a city and ultimately creates a distinctive urban identity in each case (Saradi
2012), thus enabling us to seek out their intentions and understand the historical
circumstances in which these texts were written (Kiousopoulou 2013, 124–130).[19]

Another interesting point in the later encomia of cities is that the reason usually given for composing them cites the author's need to express his gratitude to the city concerned in words. A typical example is the aforementioned praise of Constantinople by Metochites, which he is at pains to emphasize he offered in return for all that his birthplace had given him in the past (Rhoby 2012, 86). Presenting the composition of an encomiastic *ekphrasis* as a gift and a city as the recipient of the encomium transmutes the latter into a living being and gives a glimpse of the reciprocal relationship that developed between a city and its inhabitants in the final years of Byzantium. As Kiousopoulou has characteristically noted: 'A birthplace had acquired other connotations besides the geographical ones, as it concentrated within it the political and cultural signifiers that defined those connected with it. Conversely, a place was defined by the value of its inhabitants, of each one separately and all together. As a result of the reciprocal relationship created between a person and their place of birth, the city became a 'fatherland,' that is, a distinguishing mark of identity, often acquiring an even stronger emotional charge.'[20]

In recent years there has been a significant turn toward looking at space in Byzantine Studies. Tonia Kiousopoulou's study of the 'invisible' cities of the late Byzantine period, from which the passage cited above comes, is an important example of this turn, in which the real space of the cities is approached through its political and social connotations so as to create a research tool for understanding the historical development of the period.[21]

At the same time new theoretical approaches coming from other academic disciplines, such as literary criticism and political geography, have contributed to this spatial turn in Byzantine Studies. One of the most interesting new suggestions, based on the work of the French philosopher and geographer Henri Lefebvre, comes from Myrto Veikou, who proposes examining and combining readings of the perceived, conceived and lived space of the Byzantines (Veikou 2016, 2019).

According to this way of thinking, archaeological and material remains of cities make up the perceived space, while texts – like the encomia – and pictorial representations of cities can reveal other aspects of how the space looked or was understood by the Byzantines, or at least some of them. Ultimately what we lack is the lived space. The city space with its people, its smells, its sounds, and the emotions it evoked; the picture of the everyday, the trivial, or even the dark side of a Byzantine city that the rhetorical texts conceal from us. I think we can look for all these by reading both the material remains and other types of texts, such as saints' Lives through the analytical tool of lived space. In contrast with the encomiastic *ekphraseis*, the authors of hagiographical texts and by extension their readers/audience are not interested in the city's spaces but in its inhabitants and their occupations and as result particularly vivid images of urban life often emerge between the lines (Saradi 2014).

In other words, whereas in the *ekphraseis* and the encomia we take a tour round idealized and in essence 'deserted' cities, in saints' Lives, if they take place in urban surroundings, it is the people of the city who play the lead role. The best-sellers of their day with a large readership, hagiographies can, depending on the circumstances, describe a multi-coloured spectrum of the locals and foreigners who constantly frequented the city's thoroughfares and thus give us rare glimpses of everyday urban life. The poor and beggars, tavern-keepers and their wives, prostitutes and theatricals, members of the circus factions all appear in realistic detail in the Lives of saints, who from time to time assuage their hunger, distribute money, and attempt to bring them back to the true path. There is no doubt that the aim of these texts was above all to propagandize the philanthropic work of the Church and yet, if one reads between the lines, lively images of cities emerge and parts of the puzzle that make up the lived space of the people of the period can be identified.

Snapshots of the everyday life of a city, such as the down-and-outs who gather in the covered arcades of the city streets, the good Christian who, in running to get to church, trips over in the mud and spits on the ground, swearing, or the saintly women who avoid mixing with the crowd, reveal dimensions and qualities of the urban space that its material remains and its symbolic or rhetorical depictions ignore or deliberately conceal from us. Of course, it is a fact that such life-like urban images have come down to us mainly through early and middle Byzantine hagiographies (Saradi 2014). On the other hand, however, other types of literature from the late period, such as satires, can sometimes fill that lacuna.[22]

In concluding I will point to just one such glimpse, hitherto little known and commented upon. It concerns an impromptu celebration with dancing and music that took place in one of the squares in Constantinople and is described in a 10th-century saint's Life. The author recounts how the leader of a group of Paphlago-nians passing through the city, on arriving in the Forum of the Ox, began to sing with no musical accompaniment, marking time by clapping his hands. His company, made up of some thirty people, responded with a lively dance, stamping their feet loudly on the ground. A travelling musician with a lyra then steps into the limelight, which delights the revellers, who carry on singing and dancing. The celebrations are reaching a climax, when a wealthy resident of the square, comes out onto his balcony and promises to pay any musician who will play with more enthusiasm, incentivized by the promise of a monetary reward. The episode closes with a moralizing comment from the saint in question, who had been present and who utterly disapproved of such exhibitions, which he considered the devil's work (Messis 2017). This episode adds the city's characters, sounds, movements, and emotions to a square in the Byzantine capital and gets us closer to the urban space as experienced by its people. It also makes us wonder how many similar 'lived spaces' could be concealed between the lines of Byzantine texts?

Acknowledgments

The paper was first presented at the conference *Art and the City,* organized by the Hellenic Open University at the Acropolis Museum in Athens (March 17–18, 2017). I would like to thank Prof. Argyro Loukaki for the invitation to participate at the conference and this volume. Many thanks to Valerie Nunn for providing the English translation.

Notes

1 On *ekphraseis*, see also Webb (2009, 2011).
2 A typical example is Libanius' *Antiochikos,* proclaimed during the Antiochean Olympic Games of 356 AD (Voudouri 2016, 205–254).
3 This particular debate relates to the so-called Madaba map (second half of the 6th century AD), that is, the large mosaic floor with images of the most important cities and sites of the Holy Land (Piccirillo and Alliata 1998). 1999
4 Weitzmann (1979, 176–177). For images of the statuettes, see also www.britishmu seum.org/research/collection_online/collection_object_details.aspx?objectId=61305 &partId=1&museumno=1866,1229.21&page=1 (29/5/2019). The motif would live on into the middle Byzantine period, as is revealed by the personifications of cities on the Joshua Roll, one of the most famous Byzantine manuscripts of the so-called Macedonian renaissance (Evans and Wixom 1997, 238–240). For images of the Joshua Roll (Vat.Pal.gr.431): https://digi.vatlib.it/view/MSS_Pal.gr.431.pt.B (Accessed 29.05.2019).
5 As for example the townscape in the Sala della Pace of the Palazzo Pubblico in Siena (1338–1340) by Ambrogio Lorenzetti (Greenstein 1988).
6 For the frescoes of the Pantanassa, see Aspra-Vardavaki and Emmanuel (2005).
7 Tanoulas (2012, 24).
8 See, for example, the scene of the Angel smiting the Assyrians before Jerusalem in the Chora Monastery in Constantinople (1320–21), in which an orant Virgin is depicted above the gate in the city walls (Der Nersessian 1975, 343–344, pl. 462–461). This iconographic detail, which points to the identification of the Virgin with Holy Sion, the inviolate city, appears with the same symbolism from as early as the 9th century in the marginal psalters (Corrigan 1992, 52, 96–98, Figs 67, 99, 100–102).
9 This passage come from a work entitled *Periegesis,* in which the author, a Constantinopolitan church official of the first half of the 14th century, who lived in Trebizond, describes his impressions on his travels to Egypt, Syria, and the Pontos (Lampsidis 1975).
10 We find the same literary image in Theodore Metochites' oration on Nicaea (Foss 1996, 176) and in a text of Nikephoros Choumnos on Thessaloniki (Kaltsogianni et al. 2002, 39–40).
11 It is interesting to note that *ekphraseis* of aristocratic residences reveal the close relationship between an individual and their home as part of their identity and a sign of their prominent social status (Tanoulas 2018).
12 A typical example is Jerusalem in the scenes of the Entry into Jerusalem in Palaiologan monuments, in which the domed building in the center is identified with the Holy Sepulchre, see Aspra-Vardavaki and Emmanuel (2005, 28, n. 163).
13 As, for example, in cod. Stavrou 109, Jerusalem, Greek Patriarchal Library (ca 1070) (Saradi 2010, Fig. 12) and in the so-called *Vatican Epithalamion* (Vat.gr. 1851), fol. 2r (https://digi.vatlib.it/view/MSS_Vat.gr.1851 2/6/2019). The latter image is all the more significant given that it is illustrating a book intended as a gift for a foreign princess, arriving in Constantinople as the bride of a Byzantine emperor (Hennessy 2017).
14 The passage comes from the description of the city by Ioannes Kaminiates in his text *On the Fall of Thessaloniki,* referring to the capture of the city by the Arabs in 904 (Kaltsogianni et al. 2002, 14).
15 In another scene in the church at Dečani (1335–1348) Demetrios is depicted using both hands to hold up one of the towers in the walls of Thessaloniki, saving it from collapse. In this pictorial convention the saint literally functions as a bastion of the city (Stojaković 1966, Fig. 2).
16 Again from the text *On the Fall of Thessaloniki* by Ioannes Kaminiates 904 (Kaltsogianni et al. 2002, 15).

17 On places of public assembly in late Byzantine cities, which – apart from the market-place – were connected mostly with churches and their precincts, see Kiousopoulou (1993).
18 For a similar interpretation of the *ekphraseis* overall, see James and Webb (1991); Foskolou (2018).
19 It is interesting to note that what the pictorial representations of cities suggest in the middle and late Byzantine period, that is, an urban landscape without town planning, is confirmed by archaeological studies of the material remains, see Bouras (2002).
20 Kiousopoulou (2013, 127).
21 See also Kiousopoulou (2011), on approaching late Byzantine Constantinople from this point of view and Kiousopoulou 2018 on Thessaloniki. It is important to note that a pioneering example of this approach is found in Konstantakopoulou's monograph on Thessaloniki (1996).
22 See, for example, the so-called comedy of Katablattas, in which on a secondary level – behind the libelling of the main character – lively aspects of the urban life of Thessaloniki emerge, Canivet and Oikonomides (1982–1983, 53ff).

Bibliography

Aspra-Vardavaki, M. and Emmanuel, M. 2005. *He Mone tes Pantanassas ston Mystra. Oi Toichografies tou 15ou Aiona. (The monastery of Pantanassa in Mistra. The Wall Paintings of the 15th Century)*. Athens: Emporiki Trapeza tes Ellados.
Bakirtzis, N. 2012. 'Ta Teiche ton Byzantinon Poleon: Aisthetike, Ideologies, Symbolismoi (The Walls of the Byzantine Cities: Aesthetics, Ideology, Symbolisms).' In Kiousopoulou, T. ed. 2012, 139–158.
Bauer, F.A. 2013. *Eine Stadt und ihr Patron. Thessaloniki und der Heilige Demetrios*. Regensburg: Schnell Verlag.
Bouras, C. 2002. 'Aspects of the Byzantine City, 8th-15th Centuries.' In Laiou, A.E. ed., *The Economic History of Byzantium: From the 7th Through the 15th-Century*, vol. II. Washington, DC: Dumbarton Oaks Studies 39, 497–528.
Bowersock, G.W. 2006. *Mosaics as History. The Near East from Late Antiquity to Islam*. Cambridge, MA and London: Harvard University Press.
Canivet, P. and Oikonomides, N. 1982–1983. '[Jean Argyropoulos], La Comédie de Katablattas: Invective Byzantine du XVe Siècle.' In *Diptycha* 3, 5–97.
Corrigan, K. 1992. *Visual Polemics in the Ninth-Century Byzantine Psalters*. Cambridge and New York: Cambridge University Press.
Ćurčić, S. and Hadjitryphonos, E. eds. 2010. *Architecture as Icon. Perception and Representation of Architecture in Byzantine Art*. Exhibition catalogue, Princeton University Art Museum, 06.03.2010–06.06.2010. Princeton, NJ: Princeton University Art Museum and Yale University Press.
Der Nersessian, S. 1975. 'Program and Iconography of the Frescoes of the Parecclesion.' In Underwood, P. ed., *The Kariye Djami*, vol. 4. *Studies in the Art of Kariye Djami and its Intellectual Background*. Princeton, NJ: Princeton University Press, 303–349.
Evans, H.C. and Wixom, W.D. eds. 1997. *The Glory of Byzantium: Art and Culture of the Middle Byzantine Era: AD 843–1261*. New York: The Metropolitan Museum of Art.
Fenster, E. 1968. *Laudes Constantinopolitanae*. Munich: Byzantina Monacensia 9.
Foskolou, V. 2018. 'Decoding Byzantine Ephraseis on Works of Art. Constantine Manasses's Description of Earth and its Audience.' In *Byzantinische Zeitschrift* 111, 71–102.
Foss, C. 1996. *Nicaea: A Byzantine Capital and its Praises*. Brookline, MA: Hellenic College Press.

Hennessy, C. 2017. 'The Vatican Epithalamion.' In Tsamakda, V. ed., *A Companion to Byzantine Illustrated Manuscripts*. Leiden and Boston: Brill, 177–182.

Hunger, H. 1978. *Die Hochsprachliche Profane Literatur der Byzantiner. Erster Band: Philosophie, Rhetorik, Epistolographie, Geschichtsschreibung, Geographie*. Munich: C.H. Beck'sche Verlagsbuchhandlung.

Gittings, E.A. 2003. 'Women as Embodiments of Civic Life.' In Kalavrezou, I. ed., *Byzantine Women and Their World*. Cambridge, MA: Harvard University Art Museum and New Haven, CT: Yale University Press, 35–65.

Greenstein, J.M. 1988. 'The Vision of Peace: Meaning and Representation in Ambrogio Lorenzetti's Sala della Pace Cityscapes.' In *Art History* 11(4), 492–510.

James, L. and Webb, R. 1991. '"To Understand Ultimate Things and Enter Secret Places": *Ekphrasis* and Art in Byzantium.' In *Art History* 14, 1–17.

Kalopissi-Verti, S. 2013. 'Mistra. A Fortified Late Byzantine Settlement.' In Albani, J. and Chalkia, E. eds., *Heaven and Earth. Cities and Countryside in Byzantine Greece*. Athens: Hellenic Ministry of Culture and Benaki Museum, 224–239.

Kaltsogianni, E., Kotzabassi, S. and Paraskevopoulou, S. 2002. *Thessaloniki in the Byzantine Literature. Rhetorical and Hagiographical Texts*. Thessaloniki: Byzantine Research Centre.

Kiousopoulou, T. 1993. 'Lieux de Communication et Villes Byzantines Tardives. Un Essai de Typologie.' In *Byzantinoslavica* 54(2), 279–287.

Kiousopoulou, T. 2011. *Emperor or Manager. Power and Political Ideology in Byzantium before 1453* (trans. Paul Magdalino). Geneva: La Pomme d'or.

Kiousopoulou, T. ed. 2012. *Oi Byzantines Poleis (8os – 15os Aionas). Prooptikes tes Ereunas kai Nees Ermeneutikes Proseggiseis. [The Byzantine Cities (8th – 15th Centuries). Research Perspectives and New Interpretive Approaches]*. Rethymnon: Publications of the School of Philosophy, University of Crete.

Kiousopoulou, T. 2013. *Oi 'Aorates' Byzantines Poleis ston Elladiko Choro (13os – 15os Aionas [The 'Ivisible' Byzantine Cities in the Greek Space (13th-15th Century)]*. Athens: Polis Editions.

Kiousopoulou, T. 2018. 'To *Bouleuterion* tes Thessalonikis (The *Bouleuterion* of Thessaloniki).' In Kontogiannopoulou, A. ed., *Cities and Power in Byzantium During the Palaeologan Era (1261–1453)*. Athens: Academy of Athens Research Center for Medieval and Modern Hellenism, 109–119.

Konstantakopoulou, A. 1996. *Byzantine Thessaloniki. Choros kai Ideologia (Byzantine Thessaloniki. Space and Ideology)*. Ioannina: Dodoni, Suppl. 62.

Lampsidis, O. 1975. *Andreou Libadenou Bios kai Erga (Life and Works of Andreas Livadenos)*. Athens: Archeion Pontou, Suppl. 7.

Maguire, H. 1993–1994. 'The Beauty of Castles: A 10th-Century Description of a Tower at Constantinople.' In *Deltion Christianikes Archeologikes Etaireias* 17, 21–24.

Messis, Ch. 2012. *Villes de Toute Beauté. L' Ekphrasis des Cités dans Les Littératures Byzantine et Byzantino-slave*. Actes du Colloque International, Prague, 25–26.11.2011. Paris : Centre d' Études Byzantines, Néo-helléniques et Sud-est Européennes, École des Hautes Études en Sciences Sociales. Dossiers Byzantins 12. 'De l'Invisible au Visible: Les Éloges de Venise dans la Littérature Byzantine.' In Odorico, P. and Messis, Ch. 2012, 149–179.

Messis, Ch. 2017. 'Mousike, Choros kai Lipare Euochia: Logotechikes Eikones tes Paphlagonias kata te Mese Byzantine Periodo (Musik, dance and Fat Meals: Literary Images of Paphlagonia in the Middle Byzantine Period).' In *Deltio Kentrou Mikrasiatikon Spoudon* 20, 63–88.

Odirico, P. and Messis, Ch. eds. 2012. *Villes de Toute Beauté. L'Ekphrasis des Cités dans Les Littératures Byzantine et Byzantino-slave*. Actes du Colloque International, Prague, 25–26.11.2011. Paris: Centre d'Études Byzantines, Néo-helléniques et Sud-est Européennes, École des Hautes Études en Sciences Sociales. Dossiers Byzantins 12.

Parani, M. 2016. 'The Joy of the Most Holy Mother of God the Hodegetria the One in Constantinople': Revisiting the Famous Representation at the Blacherna Monastery, Arta.' In Gerstel, S.E.J. ed., *Viewing Greece: Cultural and Political Agency in the Medieval and Early Modern Mediterranean*. Turnhout: Brepols Publishers, 113–145.

Pernot, L. 1993. *La Rhétorique de l'Éloge dans le Monde Gréco-romain*, vols I–II. Paris: Institut d'Études Augustiniennes.

Piccirillo, M. and Alliata, E. eds. 1999. *The Madaba Map Centenary 1897–1997. Travelling Through the Byzantine Umayyad Period*. Proceedings of the International Conference held at Amman, 7–9.04.1997. Jerusalem: Studium Biblicum Franciscanum Collectio Maior, 40.

Politis, I. ed. 2013. *Theodoros Metochites, Byzantios, e peri tes Basilidos Megalopoleos. Kosmología kai Retorike kata ton XIV Aiona (Theodoros Metochites, Byzantios or about the Imperial Megalopolis. Cosmology and Rhetoric in the 14th Century)*. Athens: Zitros Editions.

Poulsen, B. 2014. 'City Personifications in Late Antiquity.' In Birk, S., Kristensen, T.M. and Poulsen, B. eds., *Using Images in Late Antiquity*. Oxford and Philadelphia: Oxbow Books, 209–226.

Rhoby, A. 2012. 'Theodoros Metochites' *Byzantios* and Other City *Encomia* of the 13th and 14th Centuries.' In Odorico, P. and Messis, Ch. 2012, 81–99.

Saradi, H. 1995. 'The Kallos of the Byzantine City: The Development of a Rhetorical Topos and Historical Reality.' In *Gesta* 34(1), 37–56.

Saradi, H. 2006. *The Byzantine City in the Sixth Century: Literary Images and Historical Reality*. Athens: Society of Messenian Archaeological Studies.

Saradi, H. 2008. 'Towns and Cities.' In Jeffreys, E., Haldon, J. and Cormack, R. eds., *The Oxford Handbook of Byzantine Studies*. Oxford and New York: Oxford University Press, 317–327.

Saradi, H. 2010. 'Space in Byzantine Thought.' In Ćurčić, S. and Hadjitryphonos, E. 2010, 73–111.

Saradi, H. 2011. 'The Monuments in the Late Byzantine *Ekphraseis* of Cities.' In Vavřínek, V., Odorico, P. and Drbal, V. 2011, 179–192.

Saradi, H. 2012a. 'Idyllic Nature and Urban Setting: An Ideological Theme with Artistic Style or Playful Self-indulgence.' In Odorico, P. and Messis, Ch. 2012, 9–36.

Saradi, H. 2012b. 'The Byzantine Cities (8th–15th Centuries): Old Approaches and New Directions.' In Kiousopoulou, T. 2012, 25–45.

Saradi, H. 2014. 'The City in Byzantine Hagiography.' In Efthymiadis, S. ed., *The Ashgate Research Companion to Byzantine Hagiography*, vol. II, Genres and Contexts. Farnham and Burlington: Ashgate, 419–452.

Stojaković, A. 1966. 'Quelques Représentations de Salonique dans la Peinture Médiévale Serbe.' In *Charisterion eis Anastasion K. Orlandon (Studies Presented to Anastasios K. Orlandos)*, vol. II. Athens: Library of the Archaeological Society at Athens no 54, 25–48.

Tanoulas, T. 2012, 'Anazetontas ten Antilepse tou Astikou Chorou sto Byzantio (In Search of the Perception of Urban Space in Byzantium).' In Kiousopoulou, T. 2012, 15–24.

Tanoulas, T. 2018. 'In Search of the Perception of Architectural Space in Byzantine Literature.' In Zambas, K., Korres, M., Mallouchou-Tuffano, F. and Mamaloukos, S. eds.,

Heros Ktistes. Mneme Charalampou Boura (Heros Ktistes. Memory of Charalampos Bouras), vol. II. Athens: Melissa Editions, 551–562.

Vasilakeris, A. 2014. 'Theatricality of Byzantine Images: Some Preliminary Thoughts.' In Öztürkmen, A. and Birg, E. eds., *Medieval and Early Modern Performances in the Eastern Mediterranean*. Turnhout: Brepols, 385–398.

Vavřínek, V., Odorico, P. and Drbal, V. eds. 2011. *Ekphasis. La Représentation des Monuments dans les Littératures Byzantine et Byzantino-slave. Réalités et Imaginaires*. Prague: *Byzantinoslavica* LXIX/3 (Supplementum).

Veikou, M. 2016. 'Space in Texts and Space as Text: A New Approach to Byzantine Spatial Notions.' In *Scandinavian Journal of Byzantine and Modern Greek Studies* 2, 143–175.

Veikou, M. 2019. 'The Reconstruction of Byzantine Lived Spaces: A Challenge for Survey Archaeology.' In Diamanti, C. and Vassiliou, A. eds., *Έν Σοφίᾳ Μαθητεύσαντες: Studies in Honour of Sophia Kalopissi-Verti*. Oxford: Archaeopress, 17–24.

Velmans, T. 1964. 'Le Rôle du Décor Architectural et la Représentation de l'Éspace dans la Peinture des Paléologues.' In *Cahiers Archéologiques* 14, 183–216.

Voudouri, A. 2016. *Aftotele Egkomia Poleon tes Ysteres Byzantines Periodou ypo to Prisma tes Progenesteres Paradoses tous (Independent Encomia of Cities of the Late Byzantine Period under the Light of their Earlier Tradition)*. Unpublished doctoral thesis. Athens: University of Athens, Department of Philology.

Webb, R. 2009. *Ekphrasis, Imagination and Persuasion in Ancient Rhetorical Theory and Practice*. Farnham and Burlington: Ashgate.

Webb, R. 2011. '*Ekphraseis* of Buildings in Byzantium: Theory and Practice.' In Vavřínek, V., Odorico, P. and Drbal, V. 2011, 20–32.

Weitzmann, K. ed. 1979. *The Age of Spirituality. Late Antique and Early Christian Art, 3rd to 7th Century*. New York: Metropolitan Museum of Art and Princeton, NJ: Princeton University Press.

4 Art and urban planning as identity and reflection of the great city

(Constantinople) and the sacred polis (Jerusalem) in the Byzantine provinces: the case of Cyprus

Charalampos G. Chotzakoglou

For many centuries Rome served as the city-model for the entire Roman Empire, with building types such as the Comitium, the Senate (Curia), the Basilica, the Hippodrome, the Baths, the Amphitheater, and the Temples of the imperial cult (Favro 1998, 34ff). Any distant Roman district was a miniature of Rome regarding its urban design, as was also Constantinople when it was designed by Constantine the Great (Bassett 2007; Koder 2018).

It has been symbolically stated that Christianity was founded on the 'three hills' of the Acropolis, Golgotha, and Capitolium. This metaphorical definition describes successfully the spiritual pillars of the Byzantine spirit and thought, as the transition from pre-Christian Antiquity to the Christian world was reflected on the form of the city, on its urban structure and on its models, which influenced the regional centers of the Byzantine Empire.

Along with the transfer of the Roman Empire's capital from Rome to Constantinople and the gradual expansion of Christianity, a further major change took place. On the one hand Constantinople became the model of urban organization, known as 'Polis' or 'Megalopolis,' since it was the largest city of the empire (Skarlatou 1851, 59–60). On the other hand, Jerusalem, known as the Holy City (Ousterhout 2006, 100; Koder 2018, 30) with its holy sites offered another model, since Christianity sprang up and influenced even Constantinople itself (Ousterhout 2006, 109), as the early proclamation of Christianity as official religion of the Byzantine Empire proved.

Even the connection of the Byzantine Emperor to the pilgrimages of the Holy Land as benefactor through the building, restoration, and protection of basilicas (Ousterhout 2006, 107) and his relation to special holy relics (e.g., Holy Cross, Hand of John the Forerunner) (Kalavrezou 1997, 57–67) included also elements for his political legitimation (Upton 2007, 73–74; Perrone 1999; McLynn 2004; Lenski 2004).

Nevertheless, this Graeco-Roman pagan world did not cease to have an effect on the formation of the new Christian one. Constantine the Great founded Constantinople, endowing it with myths and traditions of the Graeco-Roman past, seeking to raise the prestige and power of the new capital through all types of

Map 3 Cyprus: Location of churches and monasteries in the text.
© Map: Ch. Chotzakoglou. Source: George Chr. Papadopoulos.

culture (Alföndi 1947; Chatzelazarou 2016, 1–49, 156–200). The emperors transferred Graeco-Roman works of art (Mango 1963; Guberti-Bassett 1991; Bakirtzis
1993; Kaldellis 2007) to Constantinople and decorated the Fora and the Hippodrome (Skarlatou 1851, 56–57, 62) transforming it into a Museum of History of
Art of the Graeco-Roman spirit, to an 'eye of the Oecumene' and 'navel of the
earth,' as it was praised in the contemporary sources (Magdalino 2005; Boeck
2017, 246).

An approach, which some centuries earlier had been proclaimed by Pericles
himself in his Funeral Oration (Epitaph, 2.36–2.42), when he was referring to
Athens: 'In short, I say that as a city we are the school of Hellas.' Characteristic is
the imperial Edict of Theodosius the Great (*Codex Theodos.* 16.10.8) addressed to
Palladius, duke of Osroene (November 30, 382) on the occasion of a pagan temple
in the city of Edessa in Syria: although the cease of sacrifices to the pagan gods
was proclaimed, the temple would remain open to the common use of the people
and everything in the temple would be estimated according to its artistic value and
not to its sanctity (Tilden 2006, 265–266; Crogiez, Jaillette, and Poinsotte 2009).
The case of the tripod of Delphi, transferred to the Hippodrome in Constantinople,
offers a link between the pre-Christian world and Christianity; imperial corona-
tions were taking place opposite the tripod and emperors thus received miraculous
powers (Baran Çelik 2018, 195–198). The tripod migrated to the sacred art as
iconographic motif, seen in its characteristic representation in the scene of the
Prayer of Anne, mother of the Virgin, or the apocryphal Annunciation of Mary,
preserved also in examples of monumental Byzantine painting in Cyprus (see

Figure 4.1 Detail of the two-headed snakes of the spring of the scene of the apocryphal
          Annunciation of the Virgin. Church of the Holy Virgin of Moutoullas (13th
          century), Cyprus.

© Photo: Ch. Chotzakoglou.

Figure 4.1), as well as in illuminated manuscripts up to the Ottoman period. Thus
the effect of the topography of the City was demonstrated in the iconographic
representation of the landscape (Stylianou and Stylianou 1996, Fig. 38; Chotzakoglou 2005b, Fig. 335–336; Chotzakoglou 2010, 434–435, Fig. 11).

These links to the antiquity remained vivid throughout the eleven or so centuries of the Byzantine Empire, as a characteristic Cypriot example demonstrates.
The representation of the Baptism of Christ in Pammakaristos-Church in Constantinople along with pagan traditions, which linked Venus' birth with Paphos
inspired a local painter during the late Byzantine period and created a unique
iconographic detail in Byzantine art, connecting the ancient topography to the
Christian world. In the scene of the baptism of Christ at St. Paraskeve church in
Geroskepou, instead of the usual representation of the personification of the Sea,
a female figure was painted, the Paphian Aphrodite emerging from a shell (see
Figure 4.2). In Kouklia near Paphos a conical pagan stone was preserved in front
of the Holy Virgin's church and was included into the local cult, processions, and
customs of the church community (Paraskevopoulou 1982; Hadjikosti and Hadjisavvas 2008). Similar cases are observed in epithets related to Venus in Cyprus

Figure 4.2 Baptism of Jesus with the representation of Venus. St. Paraskeve church in Geroskepou, Cyprus.

© Photo: Ch. Chotzakoglou.

(e.g., Morpho, Oreia), which were adjusted in a Christian environment: the town with the Temple of 'Aphrodite Morpho' became the Byzantine town 'Theomorphou' (today Morphou). The church on the top of the hill, where a pagan sanctuary of 'Aphrodite Oreia' was once located, is known since at least the medieval period under the name 'St. George Oreiates' (Chotzakoglou 2013). In this way not only the ancient topography remained vivid, but also characteristics related to Venus were absorbed by the Christian religion and the sanctity of the place was preserved in public memory.

The center of Constantinople was divided by *Messe Hodos* (the Roman Road of Triumphs), which led to the Great Church. Hagia Sophia, the seat of the Archbishop of New Rome and Oecumenical Patriarch was near the administrative center of the empire, next to the Palace (Restle 1976, 23) and in this way the City was the synthesis of historical, religious and political traditions, which were carried out by the centers of power through architecture and art. At least 21 regional capitals of the Byzantine Empire, including Lefkosia in Cyprus (Tsames 1983,

73) have been recorded with their central road dividing the city and leading to the Cathedral, which was also dedicated to Hagia Sophia (Holy Wisdom of God) (Kalokyris 2004; Magdalino 2005; Chrysos 2016/7, 1017).

The building of monasteries, which gradually gained special importance, was linked to special processions and through their official ritual monasteries were connected to the emperor, and transmitted to the citizens the elements of the two fundaments of the empire: Christianity and imperial authority (Baldovin 1987, 167–226; Boeck 2017, 256). Reference of the sources to the participation of the emperor and the patriarch in several processions and ritual ceremonies of the imperial family, the clergy and the palace officials is characteristic. Probably one of the most notable processions was that of the icon of the Virgin of Hodegon, the palladium of Constantinople (Angelidi and Papamastorakis 2000, 373–387) painted by Luke the Evangelist, or the procession of *adventus imperatoris* (imperial arrival), when the Emperor on his way back from expeditions was entering Constantinople through the Golden Gate and was praying in the Acheiropoietos Church (Church of the Virgin of the Abraamites) (Skarlatou 1851, 329–332; De Vet and Mayer 2019, 314). It was in this way that the public squares were transformed into 'theaters of memory,' where the racial, historical, religious, and mythological identities of the city[1] were portrayed (Memluk 2013; Yates 1984).

The names of several monasteries of the Byzantine capital and its surrounding areas were imitated in regional centers, creating *loca sancta*. The case of Cyprus is characteristic, where monasteries bear names of Constantinopolitan and Asia Minor convents: Acheiropoietos, Mangana, Styloi, Symboulos, Katharon, Agros, Antiphonetes, Phaneromene (Chotzakoglou 2005a, 515–516; Triantaphyllopoulos 2010).

In a similar way, not only homonymous pilgrimages were created in the Byzantine periphery, but their Governing Orders were often inspired by the *Typicon* and the ritual order of the metropolitan foundation (Ousterhout 2006, 106–107). The Kykkos monastery was founded in Cyprus according to the ritual of the Constantinopolitan Hodegon monastery, imitating the ritual of the procession of the holy icon in Constantinople and the established fasting (Chotzakoglou 2009, 44–45). Additionally, the huge number of processions in the Constantinopolitan monasteries was imitated in the periphery, as is shown both in the sources and through the preserved processional crosses (Chotzakoglou 2005a, 732–735, Fig. 761–767) and double-icons (Chotzakoglou 2005a, 659–667, Fig. 570, 573).

The 'usual miracle,' the miraculous lifting of the veil of the holy Virgin's icon at the monastery of Blachernae in Constantinople every Friday evening (Grumel 1931) seems to have influenced a similar ritual regarding another St. Luke's icon in Attaleia. Similarly, another 'usual miracle' was taking place every Tuesday in Constantinople resulting to the spinning of the icon of the Virgin Mary (Lidov 2006b). Both cases find parallels in Cyprus, where several Holy Virgin's icons bear till today a veil, while Kykkos' monastery palladium (holy safeguard) has been covered with a veil since the middle Byzantine period, according to the tradition (Chotzakoglou 2005a, Fig. 566; Paliouras 2001, 89; Chotzakoglou 2009, 45); in a similar way the 'usual miracle' of the Good Thief's Cross in Cyprus consisted in its swinging in the eastern cave of the Stavrovouni convent (Stavrovouniou Monastery 1998, 112).

Likewise, the topography of the Holy Land did not influence only regional centers of the Byzantine Empire and the West (Kalavrezou 1997, 55–57; Bacci 2003; Ousterhout 2006, 100–101; Piccirillo 2009; Lidov 2012), but also Constantinople. The imperial efforts to translate ca. 3,600 relics to the Byzantine capital, as well as heirlooms related to prophets, apostles, and saints[2] (Wortley 1982; Ousterhout 2006, 102–103), and holy objects related to Jesus himself (e.g., the Holy Cross), as well as objects not created by human hands (e.g., Mandelion, Keramion) (Cameron 1984; Kalavrezou 1997, 55–57) demonstrate the potential of Constantinople to have pilgrimages equally famous as those in Jerusalem (Thomov 2013), as confirmed by Nicholaos Mesarites (Heisenberg 1922).[3] In this framework, the Apostoleion was built in order to host relics of Apostles (e.g., Andrew, Timothy, Luke), and the Church of the Holy Virgin Couratoros to preserve the relics of Sts Mary and Martha, sisters of St. Lazarus (Ousterhout 2006, 107). Holy relics of well-known saints from Cyprus followed the same way to Constantinople.[4] The relics of St Lazarus were translated from Larnaka in a sarcophagus by order of the Emperor Leo VI to Hagia Sophia and were later placed in the newly erected St. Lazarus church (Jenkins et al. 1954, 20–25; Chotzakoglou 2004). In Cyprus several holy relics are recorded in mainly post-Byzantine testimonies of pilgrims.[5] Furthermore, miraculously preserved traces of hands, knees, feet, and so forth of holy persons in Palestine launched the appearance of similar divine prints all over the Christian world. Several of them are located in Cyprus, related to the Holy Virgin, as well as other saints[6] (Dunbabin 1990; Ioannou 2007; Popović 2008; Papadopoulos and Papadopoulos 2009; Petridou 2015).

At the same time, every regional city, inspired by these great centers of the Christian world (Constantinople, Jerusalem) shaped its historical topography creating also a *locus sanctus* demonstrating local pilgrimages (Lidov 2006b, 43). In the West, Bologna became the main center, where holy relics related to Jerusalem were collected (Ousterhout 2006, 100–101; Piccirillo 2009). Sainte Chapelle was built in 1248 by King Louis IX in Paris in order to host precious holy relics, imitating the Pharos Church in Constantinople, where relics related to the Holy Passion were presented.[7] The early Christian capital of Cyprus, Constantia, which succeeded Salamis, created a holy topography including the seaside basilica of St. Epiphanius with several architectural links to the Holy Land (Chotzakoglou 2005a, 474–475), which was erected with building material from Constantinople. The Basilica of Campanopetra was built next to that, where it was probably preserved a piece of the Holy Cross (Roux 1998) along with a richly decorated Baptistery, apparently for baptisms of pilgrims, where a three-domed basilica was later built (Chotzakoglou 2005a, 498–513). From there on the pilgrims were passing through the pre-Christian tomb shown as the burial crypt of St. Catherine in Egkome near Salamis and heading to the Monastery of Apostle Barnabas, founder of the Church of Cyprus with the impressive three-aisled basilica crowned by three domes (Chotzakoglou et al. 2008, 118–121). Gradually holy places related to saints of Syropalestine and the traces they left appeared also in Cyprus (e.g., the prison of St. Paul in Paphos, the spring revealed by Apostle Andrew in the homonymous cape of Carpasia, the classroom of St. Catherine in Famagusta, the cave of Prophet Elias in Pissouri) (Chotzakoglou 2010b).

In another port of Cyprus, in Larnaca, the pilgrims approaching the shore were impressed by the three-aisled and three-domed basilica of St. Lazarus (Chotzakoglou 2004, 38–42; Chotzakoglou et al. 2008, 121–124), where a marble sarcophagus related to the saint was preserved in the burial crypt. From there the pilgrims were heading to the Monastery of the Holy Cross, known as Stavrovouni (Stavrovouniou Monastery 1998).

On the ruins of the pagan temple of Venus in Jerusalem St. Helen founded with funds of her son, Constantine the Great, the Basilica of the Resurrection (Krinsky 1970; Ousterhout 1990; Foskolou 2004), where the swinging Holy Cross was kept. In a similar way, according to the local Cypriot tradition, St. Helen also founded the Monastery of Stavrovouni on the fundaments of a pagan temple on the peak of the mountain (Stavrovouniou Monastery 1998, 23), where she dedicated the Cross of the Good Thief, which was also miraculously swinging, according to medieval travelers' texts.

Furthermore she sent Holy Nails to Constantine the Great, who fixed them on his crown and to the bridle of his horse. Sailing to Cyprus, according to local Cypriot traditions, and in order to calm the sea-storm, she threw the rest of the Holy Nails into the sea of Cyprus and managed to reach the shore of the island. On her sight the branches of the trees on both sides of her way to the inland were bending to the ground, imitating a similar situation found in apocryphal texts related to Jesus and his entrance into Jerusalem (Nikodemos 2005, 108). Having no more Holy Nails, she dedicated to the Monastery of the Holy Cross in Omodos a piece of the Holy Rope, which bound Jesus' hands on his way to the Cross on the Cavalry (see Figure 4.3).

In a similar way that names of Constantinopolitan or Asia Minor monasteries and their Typica were used as model for Cypriot convents, we trace the strong influence of the topography of Jerusalem on Cyprus. The Hermitage of Holy Neophytos in Tala imitated the atrium of the Basilica of Resurrection in Jerusalem (Chotzakoglou 2009, 922–923) and its upper part was called 'Holy Sion' (Chotzakoglou 2005a, 611–622; Kühnel et al. 2014; Kühnel 2016), a name used for other churches outside Jerusalem[8] (Harrison 1963). The toponym of 'Machairas,' today in Jordan, where the Palace of Herodes (Vörös 2015) was located, was 'transferred' to Cyprus, where Palestinian monks founded the Monastery of the Holy Virgin of Machairas (Neilos 2001). In the Pentadaktylos Mountains the monastery of St. John Chrysostomos, Patriarch of Constantinople was founded, bearing the puzzling name 'Koutzobentes.' The toponymic similarity between 'Koutzobentes' and the monastery of St. George 'Khoziba' (or Cuziba) in Jerusalem (Constantine 2014), which was founded by the Cypriot monk George (Constantine 2017) becomes even more plausible by the strong influence of Palestinian texts on its Typikon.

Figure 4.3 Detail of the Cross with the Holy Rope. Monastery of the Holy Cross in Omodos, Cyprus.

© Photo: Ch. Chotzakoglou.

The preserved post of flagellation (see Figure 4.4), where Apostle Paul was flogged, became a significant landmark of Paphos (Tsiknopoulou 1971, 31–35), imitating St. Paul's pillar in Ephesos, and reminding at the same time a similar one, Jesus' column of flagellation, the three parts of which are preserved in Constantinople (Scarlatou 1851, A 571), in Rome and in Jerusalem (Mackie 1989; Caperna 1999).

Several topographical landmarks of the Holy Land entered the Byzantine art, such as the column in Jordan marking the spot (Drosogianne 1982, 98) or the stone, where Jesus was baptized (Ristow 1965, 55–58), the representation of the walls of Jerusalem and the Basilica of Resurrection in the scene of the Entry to Jerusalem (Stavropoulou 2003), the placement of Golgotha outside the walls of Jerusalem in the scene of the Crucifixion, or the representation of the Basilica of Nazareth built on the ruins of Mary's house (Gounari 1980, 36).

In conclusion, topographical landmarks through Byzantine art were pointing to both, the imperial capital and Hagia Polis, imprinting through their visual and symbolic presence the historical, religious, and political traditions which defined

Figure 4.4 The preserved post of flagellation where Apostle Paul was flogged in Pafos.
© Photo: Ch. Chotzakoglou.

the relation between State and Church throughout the Byzantine Empire. Con-stantinople and Jerusalem became the models of the medieval Christian world and their reflections on the urban planning of the provincial towns were underlin-ing their Christian identity and their bonds to the greatest and holiest cities of the Byzantine world.

Notes

1 Under 'racial' we mean the reference of the Byzantines to the ancient Greco-Roman past, as there was no racial strife in the form we know it today.
2 For example the horn of David's anointing, the garments of Prophet Elijah, the trumpets of Jericho, the Tables of the Law.
3 The doge Pietro Centranico (1026–1031) impressed by the *mirabilia urbis* of Constan-tinople decided to steal the body of holy Sabbas and translate it to his homeland, so that Venice could imitate the imperial capital as holy center (Sanudo 1900).
4 As for example the bodies of St. Spyridon, St. Therapon, or St. Lazarus.
5 For instance, the hydria of Cana, the forearm and the lance of St. George, drops of the Holy Blood, one of Judas' pieces of silver, the Holy Nails of Jesus (Cobham 1908, 27, 30, 70, 72).
6 As for example St. George or Sta Mavra.

7 The Crown of Thorns, drops of the Holy Blood, the Holy Sponge, Lance, Nails (Kalavrezou 1997; Bacci 2003; Lidov 2012).
8 As for example Sion in Myra or a similar construction in the main church of St. Antony
monastery in Marpessa, Naxos.

Bibliography

Alföndi, A. 1947. 'On the Foundation of Constantinople: A Few Notes.' In *The Journal of
Roman Studies* 37, 10–16.

Angelidi, C. and Papamastorakis, T. 2000. 'He Mone ton Hodegon kai he Latreia tes
Theotokou Hodegetrias (The Hodegon Monastery and the Worship of the Holy Virgin
Hodegetria).' In Vasilaki, M. ed., *Meter Theou. Apeikoniseis tes Panagias ste Byzantine
Techne (Representations of the Holy Virgin in Byzantine Art)*. Athens: Benaki Museum,
373–387.

Bacci, M. 2003. 'Relics of the Pharos Chapel: A View from the Latin West.' In Lidov, A.
ed. 2003, 234–246.

Bakirtzis, C. 1993. 'He Despoina tes Thessalonikis kai he Geitonia tou Phasola.' In
Bakirtzis, C. ed., *Peza Keimena me Titlo Archaiologikai Meletai (Prose Texts Entitled
Archaeological Studies)*. Athens: Agra, 79–92.

Baldovin, J. 1987. *The Urban Character of Christian Worship*. Rome: Orientalia Christi-
ana Analecta, 228.

Baran Çelik, G. 2018. 'Entry: Serpent Head.' In Pitarakis, B. ed., *Life is Short, Art is Long.
The Art of Healing in Byzantium*. Istanbul: Istanbul Arastirmalari Enstitüsü, 195–198.

Bassett, S. 2007. *The Urban Image of Late Antique Constantinople*. Cambridge: Cambridge University Press.

Boeck, E. 2017. 'The Power of Amusement and the Amusement of Power: The Princely
Frescoes of St. Sophia, Kiev, and their Connections to the Byzantine World.' In Alexiou,
M. and Cairns, D. eds., *Greek Laughter and Tears*. Edinburgh: Edinburgh University
Press, 243–262.

Cameron, A. 1984. 'The History of the Image of Edessa: The Telling of a Story.' In Mango,
C. and Pritsak, O. eds., *Okeanos, Essays Presented to Ihor Sevcenko on his Sixtieth
Birthday*. Cambridge, MA: Ukrainian Research Institute and Harvard University, 80–94.

Caperna, M. 1999. *La Basilica di Santa Prassede*. Rome: Monaci Benedettini Vallombrosani.

Chatzelazarou, D. 2016. *He Basileios Stoa kai he Synthese tou Mnemeiakou Kentrou tes
Contantinoupoles (The Basileios Stoa and the creation of the Mmonumental Centre of
Constantinople)*. Unpublished doctoral thesis. Athens, University of Athens.

Chotzakoglou, C. 2004. *St. Lazarus in Larnaka. History, Architecture and Art*. Lefkosia:
Voula Kokkinou Editions.

Chotzakoglou, C. 2005a. 'Byzantine architektonike kai techne sten Kypro' (Byzantine
Architecture and Art in Cyprus).' In Papadopoullos, T. ed., *Historia tes Kyprou. Byzan-
tine Kypros (History of Cyprus, Byzantine Cyprus)*. Lefkosia: Archbishop Makarios III.
Foundation, Tables III.1 465–787 and III. 2.

Chotzakoglou, C. 2005b. 'He Entoichia Mnemeiake Diakosmese stous Naous tes Karpa-
sias (4th – 15th Century) (The Monumental Decoration in the Churches of Carpasia).'
In Papageorgiou, P. ed., *Karpasia*, Praktika tou I. Epistemonikou Synedriou. Lemesos:
Free Karpasia Society, 421–460.

Chotzakoglou, C. 2009. 'The Holy Virgin of Kykkos. Exploring the Transfigurations of
the Icons and its Symbolic Meaning through the Centuries.' In Guida, M.K. ed., *La
Madonna delle Vittorie a Piazza Armerina*. Napoli: Editopera Antonio Cristadi, 43–50.

Chotzakoglou, C. 2010a. 'Sholia sten Oikodomese kai ston Toichographiko Diakosmo tou Sygkrotematos tes Engkleistras tou Hosiou Neophytou sten Tala tes Paphou (Remarks on the Construction and the Pictorial Decoration of the Complex of the Hermitage of Holy Neophytos in Tala, Paphos).' In Gioultses, B. ed., *Praktika tou A' Diethnous Synedriou Ag. Neofytos o Egkleistos (Proceedings of the I' International Congress St. Neophytos The Recluse. History-Theology-Culture)*. Paphos: St. Neophytos Monastery, 919–957.

Chotzakoglou, C. 2010b. 'Entry: Apostolou Andrea Mones: Karpasia (Monastery of Apostle Andrew in Karpasia).' In *Megale Orthodoxe Christianike Egkyklopaideia* 2, 426–427.

Chotzakoglou, C. 2013. 'Entry: St. George Oreiates, Cyprus.' In *Megale Orthodoxe Christianike Egkyklopaideia* 5, 188.

Chotzakoglou, C., Chrysochou, N., Floridou, S., Foulias, A., Ierides, V., Karyda, E., Matsoukas, K., Panayiotidis, C., Pissaridis, C., Procopiou, E., Theodorou, C., Theodosiou, A. and Xenofontos, M. 2008. 'The Byzantium-Early Islam Project: The Cypriot Partnership.' In Assimakopoulou-Atsaka, P., Papakyriakou, C. and Pliota, A. eds., *Byzantio-Proimo Islam (Byzantium-Early Islam)*. Thessaloniki: Center of Byzantine Studies, 107–126.

Chrysos, E. 2016/2017. 'To Byzantio: He Autokratoria tes Kontantinoupoleos (Byzantium: The Empire of Constantinople).' In *Cypriot Studies* 78–79(3), 1005–1022.

Cobham, C. 1908. *Excerpta Cypria*. Cambridge: Cambridge University Press.

Constantine, F. 2014. *He Lavra tou Hozeva (The Hhozeva Monastery)*. Thessaloniki: Author.

Constantine, F. 2017. *Tou Hozeva Ktetores (Founders of Hhozeva)*. Athens: Author.

Crogiez, S., Jaillette, P. and Poinsotte, J.M. eds. 2009. *Codex Theodosianus Liber V. Latin Text Following the Edition of Mommsen*. Turnhout: Brepols.

De Vet, C. and Mayer, W. 2019. *Revisioning John Chrysostom. Critical Approaches to Early Christianity*. Leiden: Brill.

Drosogianne, P. 1982. *Sholia stes Toichografies tes Ekklesias tou Hagiou Ioannou Prodromou ste Megale Kastania Manes (Remarks on the Frescoes of St. John the Forerunner in Megale Kastania Manes)*. Athens: Library of the Archaeological Society in Athens, no. 98.

Dunbabin, K. 1990. 'Ipsa Deae Vestigial . . . Footprints Divine and Human on Graeco-Roman Monuments.' In *Journal of Roman Archaeology* 3, 85–109.

Favro, D. 1998. *The Urban Image of Augustan Rome*. Cambridge: Cambridge University Press.

Foskolou, V. 2004. 'Apeikoniseis tou Panagiou Tafou kai hoi Symbolikes tou Proektaseis kata ten Hystere Byzantine Periodo (Representations of the Holy Sepulcher and their Symbolic References during the late Byzantine Period).' In *Deltion tes Christianikes Archaeologikes Hetaireias* 25, 225–236.

Gounaris, G. 1980. *Hoi Toihographies ton Hagion Apostolon kai tes Panagias Rhasiotisses sten Kastoria (The Frescoes of the Holy Apostles and the Holy Virgin Rasiotissa-church in Kastoria)*. Thessaloniki: Society of Macedonian Studies.

Grig, L. 2012. 'Competing Capitals, Competing Representations: Late Antique Cityscapes in Words and Pictures.' In Grig, L. and Gavin, K. eds., *Two Romes*. Oxford: Oxford University Press, 31–52.

Grumel, V. 1931. 'Le 'Miracle Habituel' de Notre-Dame des Blachernes à Constantinople.' In *Échos d'Orient* 30, 129–146.

Guberti-Bassett, S. 1991. 'Antiquities in the Hippodrome of Constantinople.' In *Dumbarton Oaks Papers* 45, 87–96.

Hadjikosti, M. and Hadjisavvas, I. 2008. 'Kyprida Aphrodite.' In Hadjisavvas, I. ed. *Geroskepou*. Lefkosia: Chr. Andreou, 1–26.

Harrison, M. 1963. 'Churches and Chapels of Central Lycia.' In *Anatolian Studies* XIII, 117–151.

Heisenberg, A. 1922. *Neue Quelle zur Geschichte des Lateinischen Kaisertums und der Kirchenunion, I: Der Epitaphios des Nikolaos Mesarites auf seinen Bruder Johannes.* Munich: Verlag der Bayerischen Akademie der Wissenschaften.

Ioannou, S. 2007. *Ho Askas sto Diaba ton Aionon (Askas Village through the Centuries).* Lefkosia: Syndesmos Filon Anaptyxis Aska.

Jenkins, R., Laourdas, B. and Mango, C. 1954. 'Nine Orations of Arethas from Cod. Marc. Gr. 524.' In *Byzantiniche Zeitschrift* 47, 1–25.

Kalavrezou, I. 1997. 'Helping Hands for the Empire: Imperial Ceremonies and the Cult of Relics at the Byzantine Court.' In Maguire, H. ed., *Byzantine Court Culture from 829 to 1204.* Cambridge, MA: Dumbarton Oaks, 57–67.

Kaldellis, A. 2007. 'Christodoros on the Statues of the Zeuxippos Bathes: A New Reading of the Ekphrasis.' In *Greek, Roman and Byzantine Studies* 47, 361–383.

Kalokyris, K. 2004. 'Hoi Naoi tes tou Theou Sophias kai he Kathierose tou Hronou Heortasmou ton (The Churches of the Holy Wisdom and the Establishment of their Celebration).' In Kalokyris, K. ed., *Techne kai Orthodoxia (Art and Orthodoxy).* Thessaloniki: University Studio Press, 81–106.

Keshman, A. 2006. 'Emblem of the Sacred Space: The Representation of Jerusalem in the Form of the Holy Sepulchre.' In Lidov, A. ed. 2006, 252–276.

Koder, J. 2018. 'Byzantion wird Konstantinupolis: Anmerkungen zu Ortswahl und Namen.' In Katsiarde-Hering, O., Papadake-Lala, A., Nicolaou, K. and Karamanolakis, B. eds., *Hellen, Rhomios, Graikos.* Athens: Publisher? 21–33.

Krinsky, C.H. 1970. 'Representations of the Temple of Jerusalem before 1500.' In *Journal of the Warburg and Courtauld Institutes* 33, 1–19.

Kühnel, B. 2016. 'Monumental Representations of the Holy Land in the Holy Roman Empire.' In Jaspert, N. and Tebruck, S. eds., *Die Kreuzbewegung im Römisch-deutschen Reich (11.-13. Jahrhundert.).* Ostfildern: Jan Thorbecke Verlag, 319–345.

Kühnel, B., Noga-Banai, G. and Vorholt, H. eds. 2014. *Visual Constructs of Jerusalem.* Turnhout: Brepols.

Lenski, N. 2004. 'Empresses in the Holy Land: The Creation of a Christian Utopia in Late Antique Palestine.' In Ellis, L. and Kinder, F.L. eds., *Travel, Communication and Geography in Late Antiquity: Sacred and Profane.* Aldershot: Ashgate, 113–124.

Lidov, A. 2004. 'The Flying Hodegetria. The Miraculous Icon as Bearer of Sacred Space.' In Thunoe, E. and Wolf, E. eds., *The Miraculous Image in the Late Middle Ages and Renaissance.* Rome: L'Erma di Bretschneider, 291–321.

Lidov, A. ed. 2006a. *Hierotopy: The Creation of Sacred Space in Byzantium and Medieval Russia.* Moscow: Indrik.

Lidov, A. 2006b. 'Hierotopy. The Creation of Sacred Spaces as a Form of Creativity and Subject of Cultural History.' In Lidov, A. ed. 2006, 32–58.

Lidov, A 2006c. 'Spatial Icons. The Miraculous Performance with the Hodegetria of Constantinople.' In Lidov, A. ed. 2006, 349–372.

Lidov, A. 2012. 'A Byzantine Jerusalem. The Imperial Pharos Chapel as the Holy Sepulchre.' In Hoffmann, A. and Wolf, G. eds., *Jerusalem as Narrative Space.* Leiden and Boston: Brill, 63–104.

Mackie, G. 1989. 'The San Zeno Chapel: A Prayer for Salvation.' In *Papers of the British School at Rome* 57, 172–199.

Magdalino, P. 2005. 'Ho Opthalmos tes Oikoumenes kai o Omphalos tes Ges. He Konstantinoupole hos Oikoumenike Proteuousa (The Eye of the Oecumene and the Navel of the Earth. Constantinople as Oecumenical Capital).' In Chrysos, E. ed., *Byzantio hos*

Oikumene (Byzantium as Oecumene). Athens: National Hellenic Research Foundation, 107–123.

Mango, C. 1963. 'Antique Statuary and the Byzantine Beholder.' In *Dumbarton Oaks Papers* 17, 55–75.

McLynn, N. 2004. 'The Transformation of Imperial Churchgoing in the 4th Century.' In Swain, S. and Edwards, M. eds., *Approaching Late Antiquity: The Transformation from Early to Late Empire*. Oxford: Oxford University Press, 236–241.

Memluk, M. 2013. 'Designing Urban Squares.' In Ozyavuz, M. ed., *Advances in Landscape Architecture*. London: IntechOpen, 513–528.

Nikodemos. 2005. *To Apokrypho Evaggelio tou Nikodemou (The Apocryphal Gospel of Nicodemus)* (Katsaros, V. ed., Mpozines, K. comment. and transl.). Thessaloniki: Zetros.

Ousterhout, R. 1990. 'The Temple, the Sepulcher, and the Martyrion of the Savior.' In *Gesta* 29(1), 44–53.

Ousterhout, R. 2006. 'Sacred Geographies and Holy Cities: Constantinople as Jerusalem.' In Lidov, A. ed. 2006, 98–116.

Paliouras, A. 2001. 'He Eikona tes Panagias tes Kykkotissas kai he Diadose tes ston Orthodoxo kosmo (The Icon of the Holy Virgin Kykotissa and its spread to the orthodox World).' In Perdikes, S. ed., *He Hiera Mone Kykkou ste Byzantine kai Metabyzantine Archaiologia kai Techne (The Holy Kykkos Monastery in Byzantine and Post-Byzantine Archaeology and Art)*, Proceedings of conference. Lefkosia: Museum of the Kykkos Monastery, 83–104.

Papadopoulos, A. and Papadopoulos, D. 2009. *Aheritou (The Village of Aheritou)*. Larnaka: Pelio Publishers.

Paraskevopoulou, M. 1982. *Researches into the Traditions of the Popular Religion Feasts of Cyprus*. Lefkosia: M. Paraskevopoulou.

Perrone, L. 1999. 'The Mystery of Judaea (Jerome, E. 46): The Holy City of Jerusalem between History and Symbol in Early Christian Thought.' In Levine, L. ed., *Jerusalem: Its Sanctity and Centrality to Judaism, Christianity, and Islam*. New York: Continuum Publishing Company, 221–239.

Petridou, G. 2015. *Divine Epiphany in Greek Literature and Culture*. Oxford: Oxford University Press.

Piccirillo, M. 2009. 'The Role of the Franciscans in the Translation of the Sacred Spaces from the Holy Land to Europe.' In Lidov, A. ed., *New Jerusalems. Hierotopy and Iconography of Sacred Spaces*. Moscow: Indrik, 363–380.

Popović, D. 2008. 'Paying Devotions to the Holy Hermit.' In Hadjitryphonos, E. ed, *Routes of Faith in the Medieval Mediterranean*. Thessaloniki: European Centre of Byzantine and Post-Byzantine Monuments, 215–226.

Restle, M. 1976. *Istanbul, Bursa, Edirne, Iznik*. Stuttgart: Philipp Reclam.

Ristow, G. 1965. *Die Taufe Christi*. Recklinghausen: Bongers.

Roux, F. 1998. *Salamine de Chypre XV. La Basilique de la Campanopétra*. Paris: De Boccard.

Sanudo, M. 1900. 'Le Vite dei Dogi di Venezia.' In Monticolo, G. ed., *Rerum Italicarum Scriptores*, vol. 22, part 4. Città di Castello: S. Lapi.

Skarlatou, D. tou Byzantiou. 1851. *He Konstantinoupolis. Perigraphe Topographike, Archaeologike kai Historike (Constantinople. Topographical, Archaeological and Historical Description)*. Athens: Andreas Koromelas.

St. Neilos, 2001. *Typike Diataxis (Formal Monastery Order)*. Athens: Stamoulis.

Stavropoulou, A. 2003. 'Apeikoniseis tes Hierousalem stes Metabyzantines Eikones me Aforme Hena Ergo tou Ioanne Apaka (Representations of Jerusalem in post-Byzantine

Icons on the Occasion of a Work of John Apakas).' In Stavropoulou, A. and Paliouras, A. eds., *Miltos Garides (1926–1996)*, vol. II. Ioannina: University of Ioannina Press, 729–740.

Stavrovouniou Monastery. 1998. *He Hiera Mone Stavrovouniou (The Holy Monastery of Stavrovouni)*. Lefkosia: Stavrovouni Monastery.

Stylianou, A. and Stylianou, J. 1996. 'He Byzantine Techne kata ten Periodo tes Frag-kokratias (1191–1570) (The Byzantine Art during the Frankish Period 1191–1570).' In Papadopoullos, T. ed., *Historia tes Kyprou, Mesaionikon Basileion, Enetokratia*, 2 vols. Lefkosia: Hidryma Archiepiskopou Makariou, vol. 2, 1229–1407.

Thomov, T. 2013. 'Once again about the Christ Passion Relics in Hagia Sophia, Constantinople.' In *Byzantinoslavica* 71(1–2), 259–277.

Tilden, P. 2006. *Religious Intolerance in the Later Roman Empire: The Evidence of the Theodosian Code*. Unpublished doctoral thesis. Exeter: University of Exeter.

Triantaphyllopoulos, D. 2010. 'Entry: Agrou Megalou Mone (Holy Monastery of Agros).' In *Megale Orthodoxe Christianike Encyclopaedia* 1, 185–186.

Tsames, D. 1983. *Philotheou Kokkinou Bios hagiou Savva tu neou (Philotheos Kokkinos. The Vita of St. Savvas the Younger)*. Thessaloniki: Aristotle University of Thessaloniki.

Tsiknopoulou, I. 1971. *Historia tes Ekklesias Paphou (History of the Church of Paphos)*. Lefkosia: Hiera Metropolis Pafou.

Upton, T.P. 2007. *Sacred Topography: Western Sermon Perceptions of Jerusalem, the Holy Sites, and Jews during the Crusades, 1095–1193*. Unpublished doctoral thesis. Denver, CO: University of Denver.

Vörös, G. 2015. *Machaerus II: The Hungarian Archaeological Mission in the Light of the American-Baptist and Italian-Franciscan Excavations and Surveys. Final Report 1968–2015*. Milano: Edizioni Terra Santa.

Wortley, J. 1982. 'Iconoclasm and Leipsanoclasm: Leo III., Constantine V. and the Relics.' In *Byzantinische Forschungen* 8, 253–279.

Yates, F. 1984. *The Art of Memory*. London, Melbourne and Henley: Ark Paperbacks.

5 Depictions of the Virgin in an 11th-century hexaptych[1] at Sinai as perception of the city of Constantinople

Dionysis Mourelatos

The Virgin Mary (in Greek: Panagia) could represent a whole city, or be perceived as a whole city in 11th-century Byzantium. Medieval societies focused on geo-political landscapes closely connected with geo-religious concepts of space (Bogdanović 2016, 97). In this chapter we will try to also trace how Constantinople, apart from physical space, was also perceived as experienced or imagined space closely directly linked to the Virgin. Constantinople is contextualized via experience, perception, and imagination. Spatial concepts were associated with topography and faith; the glorious religious buildings of the city were linked through ceremonies performed within the city. Christian God and the Virgin Mary were even perceived as designers of Constantinople (Bogdanović 2016, 99).

All this, plus the manner in which an educated monk of Sinai Monastery perceived Constantinople in the 11th century AD, will be attested visually through a unique icon of the Sinai monastery. More specifically, the chapter will focus on a hexaptych from the monastery of Sinai (Galavaris 2009) (Figure 5.1). This hexaptych consists of four 'calendar' panels,[2] plus a panel that represented the Last Judgment[3] and a panel (Figure 5.2) with scenes of the Passion and five representations of Virgin Mary (Virgin Mary in the center, Blachernitissa, Hodegetria, Hagiosoritissa, and Chemevti) on its top (Figure 5.3).[4]

Under the central figure of the Virgin in this panel, the donor figure represented has been identified as monk John, who is mentioned in the epigram on the back of the hexaptych. This is probably the Georgian monk John (he is identified as Ioannes Tohabi), since there are bilingual (Greek and Georgian) inscriptions in all the scenes of the hexaptych as well as epigrams in Greek on the back of the panels.[5] The central figure of Virgin Mary is entitled 'Mother of God' in Greek, but the Georgian inscription reads '[Image] of the conch of St. Sophia' (Skhirtladze 2014, 378). The hexaptych can be dated with considerable certainty to the second half of the 11th century.[6] Both portraits[7] and epigrams[8] add to the complexity of these sophisticated panels. Especially the five representations of Virgin Mary are a unique composition indicating a clear connection to Constantinople, as the names of Virgin Mary that are used connect at least three of the representations to Marian shrines of Constantinople: Blachernae, Hodegon, and Hagia Soros.

Figure 5.1 The 11th-century discernible of St. Catherine's Monastery at Mount Sinai.
Source: © St. Catherine's Monastery at Mount Sinai.

Constantinople was transformed into New Jerusalem or a Second Jerusalem at least as early as the 5th century (see also Chotzakoglou's chapter here), because of the Passion relics and the relics associated with the Virgin (Talbot 2002, 60). By the early 7th century the dedication of Constantinople to the Virgin[9] was established and the city became 'Theotokoupolis' (Theotokos means Mother of God in Greek). Pilgrims were attracted to Constantinople because of the relics housed in the city's churches (Majeska 2002, 108). All these shrines were part of the itineraries of the pilgrims, probably following probably specific guides. (Majeska 2002, 104–108).

By the 9th century, after Constantinople's siege by the Russians in 860, the fame of Virgin Mary as protector of the city was well established (Mango 2000, 22). Patriarch Photios describes this siege and attributes the victory of the Byzantines to the miraculous power of the Robe (maphorion) of Virgin Mary.[10] According to John Skytitzes, the Robe secured a peace treaty with the king Symeon of Bulgaria, who sieged Constantinople in 926 (Pentcheva 2006, 52–53). The Robe of Virgin Mary that was kept in the church of Blachernae became therefore the most important Marian relic in Constantinople (Wortley 1977, 111–115). The girdle (zone) of Theotokos, another important Marian relic, was related with Theotokos Chalcoprateia church (Wortley 2005, 174). Hagia Soros (the reliquary that contained the relic) was preserved there (Zervou-Tognazzi 1986, 225, 237). It seems that the Robe gradually usurped the role of Girdle as Palladium of Constantinople, but in the texts it's not clearly distinguished from each other; two pilgrims give conflicting accounts about their location. Legends, the synaxaria,[11] and all the other texts mention the Soros of the Robe, also naming 'Soros,' after the relic that was housed in it, the chapel at Blachernae in which the Soros was preserved at least from the 10th century (Wortley 2005, 180).[12]

The church of Blachernae was clearly the most important Marian shrine in Constantinople but it was not the only Marian church where the feasts of Virgin Mary were officially celebrated;[13] it only hosted the Presentation and Dormition, whereas the other two feasts, the Annunciation and Nativity, were celebrated in the church of Virgin Mary at Chalkoprateia near Hagia Sophia (Krausmüller 2011, 226). At the church of Virgin Mary at Blachernae every Friday a procession

Figure 5.2 Panel 'a' of the Sinaitic hexaptych (detail of Figure 5.1) with scenes of the Passion and five representations of Virgin Mary.

Source: © St. Catherine's Monastery at Mount Sinai.

Figure 5.3 The five representations of Virgin Mary of panel 'a' (detail of Figure 5.2).

Source: © St. Catherine's Monastery at Mount Sinai.

started, ending at the church of Chalkoprateia, linking the two major Marian shrines of Constantinople.[14] Probably the two churches of Virgin Mary, Chalko-prateia[15] and Blachernae, functioned as two complementary foci of the Marian cult in Constantinople (Krausmüller 2011, 228).

A Latin pilgrim described a weekly 'usual miracle' the veil covering the icon of the Virgin in the church of Blachernae moved up slowly every Friday after-noon. He noted that this miracle resembles the ceremony of Good Saturday in Jerusalem (Pentcheva 2006, 159–160), enforcing the identity of the city as second Jerusalem. At least since the mid-11th century the public role of the Virgin Hodegetria as miraculous icon was likely constructed; the legend that this icon was painted by St. Luke and linked to the victory of the Byzantines over the Avars and the Persians during Constantinople's siege of 626 was wide-spread, at least during this period.[16] The earliest surviving evidence connect-ing this icon to the victory of 626 that imbued this object with special power is mentioned in a Latin text known as Anonymous Tarragonenesis, dated to 1075–98/99 (Pentcheva 2006, 56). The icon of Virgin Hodegetria became the most prominent icon-relic in the public life of Constantinople. The Hodegetria was carried each Tuesday from the Hodegon monastery through the streets of the city and placed at the altar of a different church for the celebration of Mass (Pentcheva 2006, 109).[17] It gradually became the Palladium of Constantinople.[18] Latin accounts describe massive public participation in this procession (Pentch-eva 2006, 126, 129, 135).

The church of the Virgin at Pharos (e.g., the Lighthouse) inside the complex of the Great Palace was a sacred destination of every pilgrim coming to Byzantine Constantinople. Its significance can only be compared with that of Hagia Sophia (Lidov 2012, 75–76). Most likely, the main icon of the Virgin at Pharos, named Oikokyra (Mistress of the House), was among the most important elements of the church space.[19] This icon may be identified as Virgin Chemevti (in Greek enam-elled), mentioned in the *Book of Ceremonies*.[20] However, we know nothing about its iconographical type. The church of the Virgin at Pharos[21] was not the only structure housing sacred relics inside the confines of the Great Palace. The location of sacred treasures there underlined the emperor's role as divinely-appointed guardian and protector (Klein 2006, 80) and continued being enriched with relics by the emperors (Bacci 2003, 238–239).

There were of course more Marian shrines in Constantinople,[22] as this city was gradually made into the city of Virgin Mary (see Mango 2000, 17–25). However, the significant processions in which Patriarchs took part, almost always started at Hagia Sophia (Berger 2002, 11), the most important and venerated religious edifice of Constantinople, judging from the descriptions of 11th- and 12th-century visitors like Mercati Anonymus, Anonymus Tarragonensis, and Antony of Novgorod (Majeska 2002, 94). Hagia Sophia was established as the utmost sacred place (Majeska 2002, 94) already during the reign of Leo VI (Leo the Wise, 886–912) due to its venerable relics, miraculous icons, special rites and narratives of miracles.[23] Mosaic icons of Christ, the Virgin Mary, and Archangel are attributed holy qualities (Lidov 2007, 163–167).

Leo VI created a separate sanctum inside Hagia Sophia containing important holy relics and miraculous icons, according to later sources. He retrieved to Hagia Sophia the icon with which St. Maria of Egypt was praying when she heard the voice of Virgin Mary. Later sources report that the icon of Virgin Mary mostly venerated was handled by the Patriarch during the services of feasts.[24] To sum up, the numerous Marian shrines in Constantinople, reflected the city's identification with the Virgin.

Let us now return to the Sinaitic Hexaptych and its five representations of the Virgin Mary. Ann Weyl Carr (2002, 77) remarks that it is 'a pilgrimage in paint, and it is hard not to read four named Virgins as pilgrimage sites themselves, replicas of miracle workers marking the major Marian pilgrimage sites in the City.' But how were they perceived by the painter of this icon? Why was it important to him?

Does the portrait (Figure 5.4) in this unique composition represent his self-portrait as pilgrim in Constantinople or as worshipper of the Virgin? The cult of the Virgin in the monastery of Sinai is well established since its founding along with the cult of Prophet Moses (Galavaris 2015, 27). Sinai had strong links to Constantinople, even after the Arab conquest of the region (Manafis 1990, 22), including a small dependency in Constantinople, housed in the Church of the Mangana (Patterson-Ševčenko 2006, 120). Is it therefore reasonable to assume that a Sinai monk of the 11th century could go on pilgrimage to Constantinople? Surely, there is no such evidence in the written sources. This was a long, risky, and difficult trip (Talbot 2008, 37–47). No such travels are documented. The placement of his portrait under the depiction of Virgin Mary, which can be identified with Hagia Sophia, makes tempting the assumption that an actual pilgrimage to Constantinople cannot be excluded. However, due to the iconographical similarities to both the Virgin of the conch of Hagia Sophia,[25] already noticed by scholars, and the Virgin in the mosaic with the emperors Justinian and Constantine over the entrance of the west façade to the southwest vestibule,[26] this representation was

Figure 5.4 Central representation of Virgin Mary with the self-portrait of the donor-painter
(detail of Figure 5.3).

Source: © St. Catherine's Monastery at Mount Sinai.

already identified as a reference to the church of Hagia Sophia itself (Mourelatos
2012, 74–75).[27] The epigram on the backside of the panel has been restored and
translated as follows (see Trahoulia 2002, 272–273):

> The humble monk Ioannes
> painted with desire these holy images
> which he gave to the famous church
> where he found everlasting grace.
> O Child, accept maternal intercession
> and grant full redemption from sins
> to the pitiable old man who asks it

The 'famous church' can only be the church of Hagia Sophia. However, the
destruction of this specific word of the epigram leaves another option open: the
word beloved instead of famous.[28] Definitely, we should not search for a practical use of this epigram which seems more like a self-reference, since it was not
expected to interact with contemporary beholders.[29] Apparently, the self-portrait
complements this epigram.

It is much more reasonable to assume that monk John was a spiritual pilgrim to the city of Theotokos, since first, he was a bilingual, educated monk, and second, some Sinai manuscripts could inspire such a composition.[30] Moreover, *The Life of the Virgin*,[31] a text ascribed to Maximus the Confessor surviving only in the Georgian language (Shoemaker 2008, 53), was probably known to this educated monk of Sinai, a monastery with a long historical presence of Georgian monks.[32] In this text both Blachernae and Chalkoprateia are mentioned but, as the previous analysis has manifested, Hagia Sophia, the monastery of Hodegon and the church of the Virgin at Pharos were the main attractions of pilgrims. Although theological dimensions have been attributed to this composition of five figures of the Virgin Mary,[33] we consider that it was probably more of a personal project (Lidova 2009, 83), as the epigram and the self-portrait also indicate, rather than a project designed to become public.

We can conclude, therefore, that the work in question represented a spiritual pilgrimage of a monk coming from a monastery whose patron saint was the Virgin (the Burning Bush was considered a prefiguration of Virgin Mary in the Old Testament) to Constantinople, city of the Virgin. John was probably familiar with the Marian shrines of Constantinople and indicated them through the figures of the Virgin. He painted himself kneeling before the figure of the Virgin in the center, representing the church of Hagia Sophia.[34] The depiction of Virgin Mary entitled 'Blachernitissa' represents the site of Blachernae; the depiction of Virgin Mary entitled 'Hodegetria,' the monastery of Hodegon. The depiction of Virgin Mary entitled 'Hagiosoritissa' should be connected to the monastery of Theotokos at Chalkoprateia and the depiction of Virgin Mary entitled 'Cheimevti' with the church of Pharos at the Great Palace.[35] Pilgrim accounts, oral narrations and religious texts have constructed an imagined space, an imagined city, the city of Virgin Mary and her shrines. These representations are likely connected to specific sites regardless of specific iconographical types of the Virgin Mary, since they collectively constitute a representation of the city of Theotokos, Constantinople, regardless of whether it was experienced during the pilgrimage of an educated monk from Sinai in the 11th century or it was imagined by him. John identified the urban space of Constantinople with the different representations of the Virgin Mary.

Notes

1 Diptychs are more common since antiquity, and we usually mean a laterally connected pair of panels; this hexaptych though is unique; It consisted of six panels hanged each other, although it was originally considered as tetraptych (four panels) and two separated panels by Sotirious (1956–1958, 121–123, 125–128). Galavaris' book (Galavaris 2009) is the most complete publication on this hexaptych. There is extended bibliography on this hexaptych; see also Sotiriou (1956–1958, 121–123, 125–128); Weitzmann (1966, 13–14); Galavaris (1990, 100); Kalopissi-Verti (1993–1994, 134–136); Ševčenko (2002, 52–53); Trahoulia (2002, 271–284);Weyl-Carr (2002, 75–92); Lidova (2009, 77–98); Lidova (2009b, 226–233); Mourelatos (2012, 74–75); Skhirtladze (2014, 369–386).

2 They are annual calendars of saints' days and feasts; panel b represents September to November, panel c December to February, panel d March to May, and panel e June to August (Galavaris 2009, 21, 141–142).

3 This is panel f of the hexaptych (Galavaris 2009, 21). More likely it reflects the iconography of the Second Coming in monumental art (Galavaris 2009, 142).

4 This is a panel of the hexaptych (Galavaris 2009, 21).

5 The design of the reverse is common to all the six panels. See in detail Lidova (2009, 81). For the epigrams see in detail Rhoby (2010, 50–57).

6 Galavaris (2009, 139–144); Weitzmann and Sotiriou attributed these panels also to the second half of the 11th century. However, Mouriki and Trahoulia suggest a later date for these panels, namely the beginning of the 12th century.

7 For the portraits and the self-portraits of the painters in Byzantium see in Kalopissi-Verti (1994) and specifically for the two portraits in this hexaptych in Kalopissi-Verti (1994, 134–136).

8 For the function of epigrams and inscriptions in works of art in Byzantium see Rhoby (2011, especially 319 and 326) and Papalexandrou (2001, 283). Papalexandrou notes that clearly the inscriptions were important for both patron (also the painter in our case) and beholder; the inscriptions indicate an interactive understanding by the contemporary audience.

9 It is of course known that a female personification of Tyche or Fortune is common to represent a city since the Hellenistic era. Constantinople, also characterized by the adjective *basilissa*, may also be translated ruling city, queen city, or queen of is traditionally represented as a female personification (Herrin 2000, 9). Such personifications of Constantinople in female form are dated from the 4th century (Herrin 2000, 10; Shelton 1985, 147–155). After Iconoclasm depictions of the Virgin Mary were adopted into art in the 9th and 10th centuries, creating new iconographic types. And from the 11th century onward it was the most dramatically and emotionally intense of these images which were eventually selected for incorporation in liturgical writings (Tsironis 2005, 99). After the 10th century, however, depictions of Mary, from wall paintings, to manuscripts to icons, began to present in a visual language what had long been a tradition in the key literary texts of Marian spirituality (Kalavrezou 2005, 105).

10 See in Laourdas (1959, 4, 45). The victory of the Byzantines was possible with the intervention of Virgin Mary.

11 What is meant here are texts like the legend of Galbius and Candidus (see note 30) and the synaxaria are church calendars with lections for each feast; see in Oxford Dictionary of Byzantium 3, 1991.

12 However, there are references in Byzantine sources that link Hagia Soros to Blachernae and connect it with the Robe (Maphorion) (Zervou-Tognazzi 1986, 266–267).

13 This is evident from the Typikon of the Great Church, a manual for the celebration of feasts and commemorations throughout the year, and from *De cerimoniis*, a handbook of imperial ritual commissioned by emperor Constantine VII, see Krausmüller (2011, 225).

14 (Pencheva 2006, 145). See also Bekker (1838–9, I, 694) and Patterson-Ševčenko (1991, 51).

15 It seems that between the early 8th and the early 10th centuries, authors with links to the patriarchate promoted three new feasts hosted by the Chalkoprateia, connected to Virgin Mary (Krausmüller 2011, 246). The clergy of the Chalkoprateia attempted to create a cult of the holy family, anticipating the later development in the west.

16 It seems though that it can be dated earlier, if we consider how the icon of Virgin Hodegetria became a model in the periphery of the Byzantine world. For instance in Italy the imitation of the Constantinopolitan icon and its ritual life seems to have occurred at an early date. It was natural to expect that even after the end of the Byzantine domination in 1071, Apulia, however was affected by Byzantine devotional practices: processions involving a Marian icon are recorded in Otranto as early as the 11th century, and we

find a sculpted copy of the Hodegetria, commissioned by the local turmarches Delterios in the 1030s or 1040s, inside a church in Trani (Bacci 2005, 325).

17 The Middle Byzantine procession with the Hodegetria took place early in the morning on Tuesdays. It passed through the Mese, the main commercial road of the city, and headed to a different church each week for the day's stational liturgy. Mostly Latin accounts describe these ceremonies (Pencheva 2006, 129).

18 In the paleologan era was established a close relationship between the emperor, the Empire and the icon of Hodegetria, see in detail in Angelidi and Papamastorakis (2005, 216–217). However, it was certainly a long process.

19 Lidov (2012, 100). The icon is mentioned in connection to the events of 1034, when Empress Zoe sent the most important relics of Pharos church to the rebellious Constantine Dalassenos as a guarantee for his safe return to Constantinople. They were the Holy Cross, the Mandylion, Christ's Letter to Abgar and the icon of Virgin Oikokyra.

20 Bacci (1998, 272–275, 2003, 241–242).

21 Antony of Novgorod and Mercati Anonymus have left lengthy narratives of the Virgin at Pharos (see Majeska 2002, 95).

22 Theotokos Pege, close to the Golden Gate is considered by the 6th century historian Prokopios as defender of the city, as equivalent of the Blachernae. Moreover, at the *Book of Ceremonies* an imperial procession is described that took place on the feast of the Ascension at Pege (Talbot 1994, 607). On that morning the representatives of the Guilds (the Blues and the Greens) chanted supplications on the Mother of God (Vogt 1935, I, 50–51). The monastery of the Theotokos tes Peges (the Virgin of the Source) is located just outside the walls of Constantinople. It was known for its spring, which had waters that effected miracles. It was also one of the most important shrines in Constantinople (Talbot 1996, 539–540) but it was mostly developed in the Paleologan era (Talbot 1994, 135–165).

23 Lidov (2007, 144–145). For more detail, see Lidov (2004, 393–432).

24 Lidov (2007, 144–147). Leo VI created this space possibly in 907. This program was consisted of the doors of Noe's Ark, two icons and mosaics, according to Lidov this program indicated an attempt for redemption, the only true path to salvation (see more in Lidov 2007, 169–71). See also in Lidov (1996, 535–536) and Janin (1953, 465–467).

25 For the mosaic, see in Mango (1997, 34) and Cormack (2000, 107–111)

26 For the mosaic, see in Mango (1997, 7) and Cormack (2000, 111–113).

27 Weyl Carr identifies it as the Monastery of Sinai itself; see in Weyl Carr (2002, 77).

28 See Rhoby (2010, 51). There is an destruction on the specific word, that can be fulfilled equally as Π[ΟΘΟΥΜΕ]ΝΩ (beloved) or Π[ΕΡΙΥΜ]ΝΩ (famous). However, the first seems much more probable. We are following here the fulfillment by Rhoby, who translates it as beloved.

29 For the debate over the use and role of inscriptions on works of art, see Rhoby (2011, 317–333 and especially 319, 321 and 326).

30 (Panagopoulos 2013, 4–5). The legends of Galbius and Candidus, one short and one long, can be found in manuscript Sinait. Gr. 491, which narrate the origin of Mary's robe at the Church of Blachernae. Moreover two manuscripts in Georgian on the Virgin Mary are preserved; in Sin.Geo N.75, John Damascene's Lesson for the Nativity of the Mother of God and in Sin.Geo N.6p on the Apparition of the Girdle of the Virgin, on the Presentation of the Virgin in the Temple and John Damascene's On the Presentation of the Virgin in the Temple, see Aleksidze, Shnidze, Khevsuriani and Kavtaria (2005, 427, 441).

31 The *Life of the Virgin* is a probably 7th-century text which constitutes an extant biography of the Virgin. Moreover it describes the translation of the Marian relics (the Robe to Blachernae and the Girdle to Chalkoprateia). See in detail Shoemaker (2008, 53–56, 61).

32 On the Georgian presence at Sinai, see Sotiriou (1956–1958, 115–132); Mouriki (1991, 39–40); Burtchuladze (2009, 234–239) and Ševčenko (2010, 243).

33 Trahulia connects it with the theological debate over the union of human and divine in Christ; see in Trahoulia (2002, 283–284).
34 Despite the different opinions on the date of the mosaic of Virgin Mary at the conch of the church, see the debate in Skhirtladze (2014). The iconographical similarity between the central figure of Virgin Mary at the Sinaitic hexaptych and the apse of Hagia Sophia in Constantinople can also be attributed to the selection of a common type, especially for the apse of a church.
35 Of course there is a debate over the iconographical types of the Virgin or the specific icons that are represented (Weyl Carr 2002, 77–81) but it seems that here these representations are connected to specific sites regardless to the specific icons, since they constitute as a whole a representation of the city of Theotokos, the city of Constantinople.

Bibliography

Aleksidze, Z., Shanidze, M., Khevsuriani, L. and Kavtaria, M. 2005. *Catalogue of Georgian Manuscripts Discovered in 1975 at St. Catherine's Monastery on Mount Sinai.* Athens: Greek Ministry of Culture-Mount Sinai Foundation.

Angelidi, C. and Papamastorakis, T. 2005. 'Picturing the Spiritual Protector: From Blachernitissa to Hodegetria.' In Vassilaki, M. ed. 2005, 209–224.

Bacci, M. 1998. 'Le Vergine Oikokyra, Signora del Grande Palazzo. Lettura di un Passo di Leone Tusco sulle Cattive Usanze dei Greci.' In *Annali della Scuola Normale Superiore di Pisa*, serie IV, 3, 261–279.

Bacci, M. 2003. 'Relics of the Pharos Chapel: A View from the Latin West.' In Lidov, A. ed., *Eastern Christian Relics.* Moscow: Progress-Tradition, 234–246.

Bacci, M. 2005. 'The Legacy of the Hodegetria: Holy Icons and Legends between East and West.' In Vassilaki, M. ed. 2005, 321–336.

Bekker, I. 1838–9. *Georgius Cedrenus. Ioannis Scylitzae ope, Σύνοψις ιστοριών.* Bonn: Academia litterarum regiae Borussicae.

Berger, A. 2002. 'Strassen und Plaetze in Konstantinope als Schauplaetze von Liturgie.' In Warland, R. ed., *Bildlichkeit und Bildorte von Liturgie.* Wiesbaden: Dr. Ludwig Reichert Verlag, 9–19.

Bogdanović, J. 2016. 'The Relational Spiritual Geopolitics of Constantinople, the Capital of the Byzantine Empire.' In Christie, J.J., Bogdanović, J. and Guzman, E. eds. 2016, 97–154.

Burchuladze, N. 2009. 'Georgian Icons at St. Catherine's Monastery on Mount Sinai: On Georgian-Byzantine Cultural Interrelations.' In *1st International Symposium of Georgian Culture: Georgian Art in the Context of European and Asian Cultures,* 21–29.06.2008, Proceedings. Tbilisi: International Initiative for Georgian Cultural Studies, 234–239.

Christie, J.J., Bogdanović, J. and Guzman, E. 2016a. 'Introduction: The Spatial Turn and Political Landscapes of Capital Cities.' In Christie, J.J., Bogdanović, J. and Guzman, E. eds. 2016, 3–24.

Christie, J.J., Bogdanović, J. and Guzman, E. eds. 2016b. *Political Landscapes of Capital Cities.* Boulder, CO: University Press of Colorado.

Cormack, R. 2000. 'The Mother of God in the Mosaics of Hagia Sophia at Constantinople.' In Vassilaki, M. ed. *Mother of God. Representations of Virgin Mary in Byzantine Art.* Milan: Skira, 107–123.

Galavaris, G. 1990. 'Icons from the 6th to the 11th Century.' In Manafis, K. ed., *Sinai: Treasures of the Monastery of St. Catherine.* Athens: Ekdotike Athinon, 91–101.

Galavaris, G. 2009. *An 11th Century Hexaptych of the St. Catherine's Monastery at Mount Sinai*. Venice and Athens: Hellenic Institute of Byzantine and Post-Byzantine Studies in Venice-Mount Sinai Foundation.

Galavaris, G. 2015. 'O Moyses sto Sina, he Hagia Vatos kai he Hagia Aikaterina (Moses on Sinai, the Burning Bush and St. Catherine).' In Galavaris, G. and Myriantheos, M. eds., *He Iera Mone Sina kai to Skevofylakio tes*. Athens: Holy Monastery of Sinai, 26–27.

Hahn, C. 1997. 'Seeing and Believing: The Construction of Sanctity in Early-Medieval Saints' Shrines.' In *Speculum* 72(4), 1079–1106.

Herrin, J. 2000. 'The Imperial Feminine in Byzantium.' In *Past & Present* 169, 3–35.

Janin, R. 1953. *La Géographie Ecclesiastique de l'Empire Byzantin. Les Eglises et les Monastères*, vol. III. Paris: Centre National de la Recherche Scientifique.

Kalavrezou, I. 2005. 'Exchanging Embrace. The Body of Salvation.' In Vassilaki, M. ed. 2005, 103–116.

Kalopissi-Verti, S. 1993–1994. 'Painters' Portraits in Byzantine Art.' *Δελτίον Χριστιανικής Αρχαιολογικής Εταιρείας*, ser. 4, 17, 129–142.

Klein, H. 2006. 'Sacred Relics and Imperial Ceremonies at the Great Palace of Constantinople.' In Bauer, F.A. ed., *Visualisierung von Herrschaft*, BYZAS 5, 79–99.

Krausmüller, D. 2011. 'Making the Most of Mary: The Cult of the Virgin in the Chalkoprateia from Late Antiquity to the 10th Century.' In Cunningham, M. and Brubaker, L. eds., *The Cult of the Mother of God in Byzantium: Texts and Images*. Burlington, VT: Ashgate, 219–246.

Laourdas, B. 1959. 'Fotiou Theologou, *Homiliae*.' In *Hellenika* 12, Appendix.

Lidov, A. 1996. 'Miracle-Working Icons in Church Decoration: On the Symbolic Programme of the Royal Doors of St. Sophia at Constantinople.' In Lidov, A. ed., *Miracle-Working Icon in Byzantium and Old Rus*. Moscow: Martis, 535–536.

Lidov, A. 2004. 'Leo the Wise and the Miraculous Icons in Hagia Sophia.' In Kountoura-Galaki, E. ed., *The Heroes of the Orthodox Church. The New Saints, 8th to 16th Century*. Athens: Ethniko Hidryma Erevnon, 393–432.

Lidov, A. 2007. 'The Creator of Sacred Space as a Phenomenon of Byzantine Culture.' In Bacci, M. ed., *L'artista a Bisanzio e nel Mondo Cristiano-orientalo*. Pisa: Editioni della Normale, 135–176.

Lidov, A. 2012. 'A Byzantine Jerusalem. The Imperial Pharos Chapel as the Holy Sepulchre.' In Hoffmann, A. and Wolf, G. eds., *Jerusalem as Narrative Space*. Leiden and Boston: Brill, 63–104.

Lidova, M. 2009a. 'The Artist's Signature in Byzantium; Six Icons by Ioannes Tohabi in Sinai Monastery (11th-12th Century).' In *Opera, Nomina, Historiae: Giornale di Cultura Artistica* 1, 77–90.

Lidova, M. 2009b. 'Creating a Liturgical Space: The Sinai Complex of Icons by Ioannes Tohabi.' In *1st International Symposium of Georgian Culture: Georgian Art in the Context of European and Asian Cultures*, 21–29.06.2008, Proceedings. Tbilisi: International Initiative for Georgian Cultural Studies, 226–233.

Majeska, G. 2002. 'Russian Pilgrims in Constantinople.' In *Dumbarton Oaks Papers* 56, 93–108.

Manafis, K. 1990. 'Introduction.' In Manafis, K. ed., *Sinai: Treasures of the Monastery of St. Catherine*. Athens: Ekdotike Athinon, 18–25.

Mango, C. 1997. *Hagia Sophia: A Vision for Empires*. Istanbul: Ertuğ & Kocabiyik Publications.

Mango, C. 2000. 'Constantinople as Theotokoupolis.' In Vassilaki, M. ed. 2005, 17–25.

Mourelatos, D. 2012. 'Byzantines Eikones-Menologia. To Hexaptycho tou 11^{ou} Aiona sto Sina.' In *Thirty-second Symposium of Byzantine and Post-byzantine Archaeology and Art.* Program and abstracts. Athens: Christian Archaeological Society, 74–75.

Mouriki, D. 1991. 'La Présence Géorgienne au Sinai d'après le Témoignage des Icônes du Monastère de Sainte-Cathérine.' In *Byzantio kai Georgia: Kallitechnikes kai Politistikes Scheseis (Byzantium and Georgia: Artistic and Cultural Relations)*, Conference Proceedings. Athens: National Hellenic Research Foundation, 39–40.

Panagopoulos, S. 2013. 'The Byzantine Traditions of the Virgin Mary's Dormition and Assumption.' In Vinzent, M. ed., *Studia Patristica LVI. Papers presented at the 16th International Conference on Patristic Studies held in Oxford 2011,* vol. 2. Leuven: Peeters Publishers.

Papalexandrou, A. 2001. 'Text in Context: Eloquent Monuments and the Byzantine Beholder.' In *Word & Image: A Journal of Verbal/Visual Enquiry* 17(3), 259–283.

Patterson-Ševčenko, N. 1991. 'Icons in the Liturgy.' In *Dumbarton Oaks Papers* 45, 45–57.

Patterson-Ševčenko, N. 2006. 'The Monastery of Mount Sinai and the Cult of St. Catherine.' In Brooks, S. T., ed., *Byzantium: Faith and Power (1261–1557): Perspectives on Late Byzantine Art and Culture*. New York: Metpublications, 118–137.

Pentcheva, B. 2006. *Icons and Power. The Mother of God in Byzantium.* University Park, PA: The Pennsylvania State University Press.

Rhoby, A. 2010. *Byzantinische Epigramme auf Ikonen und Objekten der Kleinkunst.* Vienna: Oesterreichische Akademie der Wissenschaften.

Rhoby, A. 2011. 'Interactive Inscriptions: Byzantine Works of Art and their Beholders.' In Lidov, A. ed., *Spatial Icons. Performativity in Byzantium and Medieval Russia.* Moscow: Indrik, 317–333.

Ševčenko, N. 2002. 'Marking Holy Time: The Byzantine Calendar Icons.' In Vassilaki, M. ed. 2005, 51–53.

Ševčenko, N. 2010. 'Manuscript Production on Mount Sinai from the 10th to the 13th Century.' In Gerstel, S. and Nelson, R. eds., *Approaching the Holy Mountain. Art and Liturgy at St Catherine's Monastery in the Sinai.* Turnhout: Brepols, 233–258.

Shelton, K. 1985. 'The Esquiline Treasure: The Nature of the Evidence.' In *American Journal of Archaeology* 89, 147–155.

Shoemaker, S. 2008. 'The Cult of Fashion: The Earliest "Life of the Virgin" and Constantinople's Marian Relics.' In *Dumbarton Oaks Papers* 62, 53–74.

Skhirtladze, Z. 2014. 'The Image of the Virgin on the Sinai Hexaptych and the Apse Mosaic of Hagia Sophia, Constantinople.' In *Dumbarton Oaks Papers* 68, 369–386.

Sotiriou, M. and Sotiriou, G. 1956–1958. *Eikones tes Mones Sina (Icons of the Sinai Monastery)*. Athens: École Française d'Athènes.

Talbot, A.M. 1994. 'Epigrams of Manuel Philes on the Theotokos tes Peges and Its Art.' In *Dumbarton Oaks Papers* 48, 135–165.

Talbot, A.M. 1996. 'Miracle-working Images at the Church of Pege in Constantinople.' In Lidov, A. ed., *Miracle-working Icon in Byzantium and Old Rus.* Moscow: Martis, 539–540.

Talbot, A.M. 2002. 'Pilgrimage in the Byzantine Empire: 7th-15th Centuries.' In *Dumbarton Oaks Papers* 56, 59–61.

Talbot, A.M. 2008. 'Pilgrimage in the Eastern Mediterranean between the 7th and 15th Centuries.' In Kazakou, M. and Skoulas, V. eds., *Egeria. Monuments of Faith in the Medieval Mediterranean.* Athens: Hellenic Ministry of Culture, 37–46.

Trahoulia, N. 2002. 'The Truth in Painting: A Refutation of heresy in a Sinai icon.' In *Jahrbuch der Oesterreichischen Byzantinistik* 52, 271–285.

Tsironis, N. 2005. 'From Poetry to Liturgy: The Cult of the Virgin in the Middle Byzantine Era.' In Vassilaki, M. ed. 2005, 91–102.

Vassilaki, M. ed. 2005. *Images of the Mother of God. Perceptions of the Theotokos in Byzantium*. Aldershot: Ashgate.

Vogt, A. ed. 1935. *Constantin VII Porphyrogénète, Le livre des Ceremonies*, Livre I, chapitres 1–46 (37). Paris: Les Belles-Lettres.

Weitzmann, K. 1966. 'Byzantine Miniature and Icon Painting in the 11th Century.' In *Thirteenth International Congress of Byzantine Studies*, Main Papers, 7. London: Oxford University Press, 207–224.

Weyl Carr, A. 2002. 'Icons and the Object of Pilgrimage in Middle Byzantine Constantinople.' In *Dumbarton Oaks Papers* 56, 75–91.

Wortley, J. 1977. 'The Oration of Theodore Syncellus (BHG 1058) and the Siege of 860.' In *Byzantine Studies/Etudes Byzantines* 4, 111–126.

Wortley, J. 2005. 'The Marian Relics at Constantinople.' In *Greek, Roman and Byzantine Studies* 45, 171–187.

Zervou-Tognazzi, I. 1986. 'L'Iconografia e la 'Vita' delle Miracolose Icone della Theotokos Brefokratoussa Blachernitissa e Odighitria.' In *Bolletino della Badia Greca di Grottaferrata* 40, 215–282.

6 Between convention and reality
Visual approaches to cities in post-Byzantine icon painting

Jenny P. Albani

In 1338–1339 the Sienese painter Ambrogio Lorenzetti (ca. 1285 – ca. 1348) painted three outstanding large-scale frescoes covering the northern, eastern, and western walls of the Room of the Nine (the Council Chamber) at the Palazzo Pubblico, the Old Town Hall of Siena. They represent the Allegory of Good Government, the Effects of Good Government on the City and the Country, and the Allegory and Effects of Bad Government (Tyranny) on the City and the Country. Beyond the fact that these paintings are the earliest medieval panoramic views of a city known so far, they also depict in detail aspects of the daily life of the citizens emphasizing the differences between a well- and an ill-governed medieval city-state. Moreover, an important innovation of Lorenzetti's frescoes in the Old Town Hall of Siena is the representation of the personifications of virtues and vices related to the governance of an Italian city-state, thus visualizing the first constitution of an Italian commune.[1]

This chapter discusses Byzantine stereotypes regarding the depiction of cities, in conjunction with Western influences, to reveal critical narratives toward living in urban centers considered 'sinful.' The meaning was the outcome of symbols and metaphors but also literary texts and preaching, alongside contemporary historical events. The Byzantine and post-Byzantine East was unfamiliar with such a naturalistic rendering of the city as Lorenzetti's as well as a visual approach to its social character. Images of cities are in most cases supplementary iconographic elements of a composition with religious content (see also Chapter 3, this volume). However, in some post-Byzantine icons, cityscapes do not serve merely as complementary motifs to religious narratives but also as visual expressions in a symbolic manner of the commissioners' notions on the moral and social character of contemporary cities. Their rendering conforms to visual conventions of Byzantine art, such as the lack of perspective, urban fabric, open spaces, and landmarks. Moreover, their buildings appear anti-realistic and on a smaller scale in comparison with human figures. Notwithstanding, through the relation of these cityscapes to the pictorial narrative, in some cases interpreted by inscriptions, criticism on post-Byzantine urban life is articulated. Case studies of this approach, which this paper aims to discuss, are three icons in Greek and Cypriot collections: the Allegory of Heavenly Jerusalem (105.5 x 82.5 x 1 cm) at the Monastery of the Most Holy Virgin Platytera, Corfu; St. Demetrios (108 x 82.5 x 2.2 cm) at the

Byzantine Museum of Antivouniotissa, Corfu (inv. no 157); the Last Judgment (83 x 119 cm), at the Byzantine Museum of the Archbishop Makarios III Foundation, Lefkosia (inv. no BM 100). Judging by their large dimensions, we may suggest that these painted panels were not intended for personal devotion, but probably placed at a public religious space. I will argue that visual approaches to the Sublime – the eternal life – in these icons invite viewers to reclaim it through repentance and ethical compliance.

Heavenly Jerusalem versus sinful Babylon

The first icon in question, dated between 1510 and mid-16th century, depicts an allegory of Heavenly Jerusalem (see Figure 6.1).[2] Numerous plethoric inscriptions against its golden background, mostly excerpts from the Book of Revelation and the Gospel of Matthew, interpret its rare subject matter. To the right and left, two cityscapes, identified by inscriptions with Holy Zion – Heavenly Jerusalem of the Scriptures – and Babylon, are the main iconographic elements.

On the right, Heavenly Jerusalem, shown in bird's eye view, is a fortified city rhomboid in form with twelve towers and twelve gateways, guarded by twelve angels.[3] It lies on the top of a steep mountain, the 'Mount of Virtues,' according to a label. One more inscription related to the holy city informs us that its fortification wall is made of precious stones and its squares paved with pure gold.[4] In the middle of this holy city, shown in the form of an enclosed garden without any buildings, Christ blesses eight choirs of the elect.

The fortified city of Babylon and its personification, a crowned young woman riding a seven-headed beast with horns, appear to our left.[5] The accompanying inscription of the city refers to Great Babylon as the mother of harlots and abominations of the earth.[6] The city is densely built but empty of people since the icon shows it in the Day of Judgment, being abandoned by its citizens, according to the Book of Revelation (Rev. 18: 4–5). A large group of people exit through the wide gate of the fortification, above which a mask motif in the form of a lion's head appears, the lion being a symbol of power but also an emblem of the Republic of Venice. They form a procession alongside a broad road toward the cave of Hell, at the right end of the image. Labels over some of these figures and inscriptions on the open scrolls of others identify 38 of them with vices: Discord, Enmity, Sorcery, Abomination, Idolatry, Fornication, Harlotry, Adultery, Hypocrisy, Heresy, Calumny, Masturbation, Malice, Sloth, Blasphemy, Injustice, Abuse, Rapacity, Babbling, Jealousy, Laughter, Hatred, Cunning, Envy, Avarice, Pride, Anger, Lust, Debauchery, Ambition, Iniquity, Self-love, Greed, Incontinence, Theft, Revelry, Gluttony, and Drunkenness.[7] Eye-catching within the procession of the ‚ personifications of vices and their fellows is the leading figure of a seductive, bare-breasted young woman standing on a golden chariot and pointing to Hell. An inscription identifies her with 'the flesh fighting against the spirit.'[8] At the entry of Hell, Death, depicted as a human skeleton with a sickle, snatches the hand of the personified Debauchery. An accompanying inscription invites the viewer to consider Death as 'the punishment of flesh and the common fate of all.'[9] Above

Figure 6.1 The Allegory of Heavenly Jerusalem. Icon. Monastery of the Most Holy Virgin
Platytera, Corfu.

Source: © Hellenic Ministry of Culture. Directorate of Archaeological Museums, Exhibitions and
Educational Programmes. Photo by Elias Eliades.

the cave, a nun, the personification of Repentance, saves a true penitent from Hell, according to a label.[10]

From the narrow gate of the wall of Babylon exit few people, most of whom are Orthodox monks and nuns. They each carry a cross on their shoulders and try to reach Heavenly Jerusalem through the rough mountainous pathway. A small group succeeds in approaching the top of the mountain, where Christ welcomes them. There are, however, two cases of monks, without a cross, who glide, stumble, and become victims of demons.

The imagery of this icon, based on the Book of Revelation, is also strongly influenced by patristic literature, mainly the Ladder of Divine Ascent by John Sinaites (before 579 – ca. 650), an edifying treatise that was very popular in monasteries of the Christian East.[11] The iconography of the pageant in our icon probably draws on Renaissance art, especially on The Triumph of Love, which forms part of the composition Trionfi, illustrating the extremely popular homonymous poem by Petrarch.[12] Moreover, West European woodcuts of the allegorical composition Dance of Death, showing skeletons – symbols of Death – clasping humans, probably inspired the figure of Death grasping the hand of the personified Debauchery in the icon.[13]

The heavily allegorical, eclectic, and prolifically annotated icon of the Platytera Monastery on Corfu is a pictorial counterpart of an Orthodox sermon. It criticizes late medieval urban life, implies Eastern asceticism as a way to salvation, and indicates the importance of repentance. It was probably commissioned for a monastery, although not for the Platytera Monastery, which was founded only in 1743. The learned blending of scriptures and visions, as well as the impact of Renaissance art on this icon, favour its attribution to a well-educated milieu in the East, with a profound knowledge of Orthodox theology on the one hand and artistic bonds with the West on the other. It is likely that the icon was painted for an Orthodox monastery on Crete, Venetian-held at the time, probably in its most important artistic hub, Chandax (Venetian Candia, present-day Herakleion).[14]

The stylistic device of antithesis, used in Greek rhetoric diachronically, also influenced the imagery of this icon to make viewers better understand its lessons: the wide gate versus the narrow one, the broad and plain road of sins in contrast to the steep and uphill paths of virtues, the crowd of sinners against the few righteous (cf. 'Many are called, but few are chosen,' Matth. 22: 14). In the Last Day, sinful Babylon features impressive buildings but no people, while Christ, angels, and the elect inhabit Heavenly Jerusalem which has no buildings made by humans.[15] The contrast between Heavenly Jerusalem and sinful Babylon in our icon is analogous to that between a well- and an ill-governed city highlighted by Lorenzetti in his Siena frescoes.

Adopting Renaissance models for the depiction of the personifications of vices walking in a solemn pageant toward Hell probably reveals a critical attitude towards urban life conforming to Western modus vivendi. Pageants were an essential element of 16th-century Venetian social life, and religious processions in which both Latin and Greek clergy had to participate were also imposed in Chandax by the Venetian colonists.[16] There is some evidence that Greek clergy refused to take part in these processions, which demonstrated its spiritual subordination to the Latin Church.[17]

Some historicist and theological interpretations of the Revelation relate symbolically sinful Babylon to Jerusalem, Rome, and Constantinople.[18] Moreover, it is worth noting that Cretan cities of the Venetian period and especially Chandax are criticized for their sins by late medieval literature.[19] A telling example is *The destruction of Crete*, an edifying poem by Manuel Sklavos, composed soon after 1508. It refers to a terrible earthquake in Chandax, on the night of May 29, exactly 55 years after the Fall of Constantinople to the Ottomans, which created the impression that the end times were approaching. The citizens considered the earthquake as a divine punishment for their sins and, terribly shocked, left their almost ruined city.[20] Within this climate of eschatological preoccupations, the poet aims to raise the consciousness of his contemporaries, calling them to repentance and relating a possible future destruction of Chandax (referred to as Crete) to the amorality of its citizens:

> And all this happened so that God may teach us
> that, unless we repent, He shall destroy us.
> He shall mutilate our bodies with the Turkish sword
> and we shall be held captive for our sins.
> If we, therefore, desire herbs to heal our wounds,
> let us fall on our knees before God in penitence,
> I implore you to denounce homosexuality
> and eschew blasphemy against God.
> Not only did God abominate sodomy,
> but also the disrespect shown towards His Mother and the Cross
> We the Cretans, all of us, have been irreverent towards
> the Baptism, the Anointing with Myrrh and the Holy Trinity.
> Let us stay away from usury, as the Prophet commands,
> or else be aware that this shall bring about the demise of Crete.[21]

A similar critical spirit with a moralizing intention permeates the imagery of our icon, probably painted in Chandax (that parallels sinful Babylon?) some years after the earthquake of 1508.

Post-Byzantine Thessaloniki: a sinful city?

The second icon, depicting St. Demetrios on horseback (see Figure 6.2), may provide us with one more example of a post-Byzantine cityscape in moralizing context. The painting has been dated, on stylistic grounds, to around 1600 and attributed to a Cretan artistic workshop.[22] St. Demetrios ([O ΑΓΙΟC Δ] HMHTPI[OC]) is shown as a young and handsome warrior, in full armor. Riding a powerful red horse, which is striding toward the left, he holds a spear and the horse's rein and turns his upper body toward the beholder, as in dialogue with him. At the upper left corner of the icon the hand of God, emerging in a segmentum coeli (segment of heaven), blesses the saint.

Figure 6.2 St. Demetrios. Icon. Byzantine Museum of Antivouniotissa, Corfu, inv. no 157.
Source: © Hellenic Ministry of Culture. Ephorate of Antiquities of Corfu.

St. Demetrios is portrayed in a flat field, in front of two mountains separating him from his home city, Thessaloniki, which appears at the far right. The city can be recognized only through the inscription Η ΣΑΛΟΝΙΚΗ on its fortification wall since its depiction is conventional, conforming to Byzantine stereotypes: in bird's eye view, fortified, densely built, and with coloured buildings. Three tower-like buildings with a cross on their top could represent bell towers of churches.

By the 6th century, the persona of St. Demetrios was closely associated with Thessaloniki as its patron saint, defending it through his posthumous miracles from sieges, attacks, and natural disasters. Three collections of medieval literary texts, the Miracles of St. Demetrios, refer to the saint's permanent presence and interest in his home city.[23] Several of his Byzantine depictions convey, moreover, similar messages.

A large mosaic panel (soon after 620) in the Basilica of Hagios Demetrios in Thessaloniki depicts St. Demetrios infra muros, between two donors of his church: John, Archbishop of Thessaloniki, and Leontios, Prefect of Illyricum. The saint puts his hands around the shoulders of his companions – the spiritual and civic leaders of the city – thus showing his benevolence not only to the two donors but also through them to Thessaloniki.[24] Furthermore, a 12th-century silver-gilt reliquary in the form of a small box, in the Vatopedi Monastery on Mount Athos, decorated with scenes from the life of St. Demetrios, features the saint on the city wall fighting against the Slavs.[25] The image on a 13th-century coin highlights also the saint's interest in his city: St. Demetrios and Manuel Angelos Komnenos Doukas, Despot of Epiros and Thessaloniki (ca. 1230 – ca. 1238), enthroned, hold together a model of the city.[26] Moreover, in wall paintings of the chapel of St. Demetrios at the Monastery of Dečani (1335–1350), the saint saves his city from starvation, restores a fallen tower of its fortification, and fights against its enemies, the Cumans.[27]

Returning to the post-Byzantine icon in question, we see Thessaloniki isolated, behind mountains separating it from St. Demetrios, enclosed in its wall, with its gate locked, while its patron saint has taken his lonely way toward the Lord. As Charalambos Bakirtzis pointed out, the tomb of the saint inside the Basilica of Hagios Demetrios in Thessaloniki was a locus sanctus for both Christians and Muslims, during the Ottoman occupation of the city,[28] confirming the presence of the saint's relics in the church, and hence his spiritual presence in the city. There were also, however, different contemporary beliefs, according to which the saint had abandoned his city and not saved it from the Ottomans due to its sins.[29] Indicative of these views is an enkomion (1666–1669) with a supplication to the saint by an anonymous cleric:

O Demetrios, Great Martyr of Christ the Lord, where are the wonders that you worked daily here in your homeland, how don't you help us now? How don't you visit us? Why, St. Demetrios, did you fail and turn your back on us? Don't you see the misfortunes, temptations and debts we have been surrounded by? Don't you see that we have been dishonoured and trampled by our enemies, ridiculed by the impious, mocked by the Saracens, loathed and derided by everybody? So, Saint, why don't you pity us, albeit a compatriot? Why don't you sympathize with us? Don't you hear the lament, don't you see our tears? Don't you ponder over the trials and tribulations we are going through?'[30]

The idea that Thessaloniki suffers when its patron saint is away or abandons it because of its sins also appears in medieval texts on its siege and capture by the Saracens in 904 though not in Byzantine art.[31] The 10th-century sermon of Nikolaos Mystikos, Patriarch of Constantinople, read at Hagia Sophia shortly after the capture of Thessaloniki, in which the patriarch asks the saint similar questions to those of the 17th-century anonymous writer, is an eloquent case in point:

> Where, Martyr Demetrios, is your invincible succor now? How could you allow your city to be sacked? Inaccessible to enemies under your patronage from the time the sun saw it first, how could it experience evils so great? How could you bear the pride of the impious deriding your holy protection? How did you endure and tolerate this?[32]

St. Demetrios answers, in the sermon, that he and God are displeased for the sins of people.

Could it be that our icon visualizes the belief that St. Demetrios left his city disappointed by its sins? If this is the case, then, outside of sinful Thessaloniki, St. Demetrios – having already acquired an ecumenical saintly status – addresses the beholder of the icon, promising salvation to those who pray to him through mediating for them to the Lord.

The burning city

The last icon of our survey, with a plethoric composition of the Last Judgment, comes from the Church of Panagia Phaneromene, in the old town of Lefkosia, and has been dated to the 16th century.[33] On the top of the representation Christ the Judge appears in Glory surrounded by the Virgin and St. John the Baptist, interceding for the remission of human sins, Archangels Michael and Gabriel, the heavenly court of the apostles, and angels. With his right hand, the Lord invites the righteous who approach him in clouds, while with his left he dismisses the sinners whom the fiery river drives to Hell, to the bottom right of the icon. As is usual in the iconography of the Second Coming, the angel rolling the scroll of Heaven, the Preparation of the Throne surrounded by two beggars, Adam and Eve kneeling in front of the throne, the Weighing of Souls, the Earth releasing the Dead, Paradise, and Hell also appear in the icon.

Particularly interesting for our study is the depiction of a city, rendered flat, in a brownish-black color, amid the flames of Hell (see Figure 6.3). This fortified city, with towers and a railed arched gate, is surrounded by demons that fly around or torment the damned. The motif is not merely an allusive reference to immoral urban life. Here either a sinful city is damned, as Sodom and Gomorrah were punished and burnt by God's fire, or becomes a place of torments in Hell enhancing the harrowing of the afterworld.

The motif of the burning city in Hell, uncommon for Byzantine and post-Byzantine infernal landscapes, may be associated with the visual vocabulary of the Dutch master Hieronymus Bosch (ca. 1450–1516) who used to criticize

Figure 6.3 Burning city. Detail of an icon with the Last Judgment. Byzantine Museum of
the Archbishop Makarios III Foundation, Lefkosia, inv. no BM 100.

Source: © Byzantine Museum of the Archbishop Makarios III Foundation, Lefkosia. Photo: Jenny
Albani.

inter alia the morals of urban life.[34] It forms part of his paradoxical landscapes
of Hell, The Haywain Triptych (1495–1516) and The Garden of Earthly Delights
(ca. 1510) in the Museo del Prado, Madrid, being typical examples.[35] Bosch spent
probably some time between 1499 and 1503 in Venice, where three of his paintings (two diptychs and a set of four panels in the Palazzo Ducale) are preserved.[36]
Therefore, we may assume that the painter of the Cypriot icon, criticizing the
society of his time, had copied the burning city in Hell either from one of Bosch's
paintings he had seen in Venice or from the numerous prints and paintings by
his followers and imitators. Among them a panel with the Vision of Tundal, in
the Museo Lazáro Galdiano, Madrid, attributed to the Flemish School, features a
burning city in Hell similar to that of the Lefkosia icon. Its subject, a dream of the
underworld, was popularized by Bosch's followers.[37]

To conclude: Although the three paintings discussed in this paper conform to
Byzantine stereotypes regarding the depiction of cities, their pictorial narrative
reveals a critical view of their commissioners toward sinful living in urban centres. Late medieval beholders, like the Byzantines, were acquainted with symbols
and metaphors whereas the literary texts and preaching, alongside contemporary
historical events, contributed to the understanding of their meaning. According
to these icons' lessons, redemption from sins and eternal life but also well-being
on earth presuppose true repentance and the interceding of the saints to the Lord.

114   *Jenny P. Albani*

Acknowledgments

I owe special thanks to Professor Argyro Loukaki, Hellenic Open University, who kindly invited me to participate in this edited volume and made comments on my text. Similarly, I am deeply indebted to Ms. Tenia Rigakou, Director of the Ephorate of Antiquities of Corfu, and Dr. Ioannes Eliades, Director of the Byzantine Museum of the Archbishop Makarios III Foundation, Lefkosia, for permitting me to study the icon of St. Demetrios at the Antivouniotissa Museum and that of the Last Judgment at the Byzantine Museum, Lefkosia, respectively. My thanks are extended to Dr. Ioanna Christoforaki, Academy of Athens, for sharing with me her thorough knowledge of icon painting, and Dr. Demetrios Doumas for editing the English text.

Notes

1 Polzer (2002, 63). On these frescoes, see also White (1993, 388–392, figures on 234–237); Dupont and Gnudi (1990, 104–105, figures on 100–103); Campbell (2001); Skinner (1999).
2 On the icon, see Vocotopoulos (1990, 19–22); Kephallonitou (1994); Chondroyannis (2010); Mastora (2012); Tsimpoukis (2013, 54); Markomichelaki (2014, 44); Vocotopoulos (2016, 180); Triantaphyllopoulos (2016); Gratziou (2016, 59); Albani (2017, forthcoming).
3 On the iconography of Heavenly Jerusalem, see Lidov (1998), with bibliography.
4 'Concerning Heavenly Jerusalem, gathering from the writings of Prophet Isaiah and Tobit and David, and the Revelation of John. Holy Zion, the city of God, Heavenly Jerusalem, whose foundations are in the Holy Mountains and its walls made of emerald and sapphire, and diamonds, and all manner of precious stones joined together with gold. Its battlements are of jasper and its gates of precious pearls and the squares are paved with pure gold featuring beauteous residences made of transparent glass, and pearls and glittering gold, and inside is the river of the water of life, clear as crystal, and lovely trees for the comfort and pleasure of the blessed people: and they, having come out of mighty Babylon through the narrow gateway and having walked along the sad path of the Lord, after much sorrow and suffering, toil and sweat shed on the hard road and for the sake of virtues, came here and having received from God a crown of glory enjoy the eternal bounties with much comfort and with never-ending joy, beholding the kingdom with glory and inhering for all time the land of the living.' Vocotopoulos (1990, 19); Kephallonitou (1994, 150–151). The inscription derives in the Bible (Is. 54:11–14; Tobit 13:15–18; Rev. 18).
5 An accompanying inscription, which is an adaptation of the 17th chapter of the Book of Revelation (Rev. 17:1–5), reads: 'The great whore that sitteth upon many waters and sitting upon a beast, holding a golden cup in her hand full of abominations and filthiness, from which drink those that dwell on earth and coveting earthy things and drunk with the wine of her fornication, never thinking that in one day her wounds will arrive, death and mourning, and plague and she will be utterly burnt with fire together with her servants who walk along the wide way.' Vocotopoulos (1990, 21); Kephallonitou (1994, 152).
6 'Babylon the great, the mother of harlots and abominations of the earth, which shall fall on the day of judgment, and shall become the habitation of devils and the hold of every foul spirit.' Vocotopoulos (1990, 20); Kephallonitou (1994, 151).
7 Cf. the fourth Homily by Photios, Patriarch of Constantinople (858–867, 877–886), mentioning vices of the citizens of Constantinople: drunkenness, gluttony, fornication,

injustice, hatred of one's brother, anger at one's neighbour, haughtiness, murder, envy, negligence, and indifference. Mango (1958, 106); Tsiaples (2014, 131). Some of them are identical to those of the icon in question.

8 'Flesh who fights against the spirit, who runs along all parts of the wide road unbridled and unchecked, along with her bittersweet deed and passions.' Vocotopoulos (1990, 21); Kephallonitou (1994, 152–153).

9 'Death, the punishment of flesh, is common for all, but he does not send true penitent to Hades.' Vocotopoulos (1990, 21); Kephallonitou (1994, 152).

10 'Repentance is a gift of God, which snatches away from the eternal fire those who truly repent, and saves them with the grace of Christ the Saviour.' Vocotopoulos (1990, 21); Kephallonitou (1994, 153).

11 Albani (2017; forthcoming).

12 Petrarch's poem *Trionfi* describes six allegorical visions: the triumphs of Love, Chastity, Death, Fame, Time, and Eternity/Divinity. On the poem, see Wilkins (1963); Finotti (2009); Eisenbichler and Iannucci (1990). Petrarchan Triumphs appear in art in six plethoric compositions with six chariots, each carrying a personification and drawn by symbolic beasts. On representations of the *Trionfi*, see Wyss (1998); Baskins, Randolph, and Musaccjio (2008); Pope-Hennesy and Christiansen (1980).

13 Albani (forthcoming).

14 Albani (2017; forthcoming).

15 On the antithesis in Byzantine religious literature and art, see Maguire (1981, 53–83; 2012, 144–152).

16 Muir (1979, 36–52); Georgopoulou (1995, 486–490).

17 Georgopoulou (1995, 489).

18 Tsimpoukis (2013, 191); Aalberts (1998, 245–259); Markomichelaki (2015, 184).

19 For example the tragedy *Ροδολίνος* (*Rodolinos*) by John Andreas Troilos referring to the capture of Memphis, probably a symbol of Rethymnon. Further examples are the poems *Της Κρήτης ο χαλασμός* (*The Destruction of Crete*) by Manuel Sklavos, dealing with the earthquake of 1508 in Chandax, and *Κρητικός πόλεμος* (*Cretan war*) by Marinos Tzanes Bouniales, which speaks of the siege and capture of Chandax by the Ottomans (1645–1669), as well as the text *Κρητική διήγηση* (*Cretan narrative*), dealing with the plague of 1592–1593 in Chandax. All these literary works imply that Venetian-held Rethymnon and Chandax suffered because of the sins of their inhabitants (Markomichelaki 2015, 179–212).

20 Markomichelaki (2014, 15).

21 Και ταύτα πάντα εγένησαν αγιά να μας διδάξει
και α δεν μετανοήσομεν, θέλει μασε πατάξει,
με το σπαθί το τούρκικο να κόψει τα κορμιά μας   145
και αιχμάλωτοι να γένομεν ογιά τα κρίματά μας.
Λοιπόν, αν θέλ'τε να 'βρομεν εις τες πληγές βοτάνια,
ας πέσομεν προς τον Θεόν όλοι με την μετάνοια,
ν' αφήσομεν, παρακαλώ, την αρσενοκοιτία
και εκ του Θεού το όνομα να λείψει η βλασφημία·   150
δεν εσιχάθηκεν ο Θεος μόνον την σοδομίαν,
της Μάνας του και του Σταυρού την τόσην ατιμίαν·
το βάπτισμα, το μύρωμα και την Αγιάν Τριάδα
εβλαστημούμαν Κρητικοί, όλη μας η ομάδα.
Την ζούραν όλοι ας φύγομεν, σαν το βοά ο Προφήτης,   155
ειταδεμή, κατέχετε, θάνατος έν' της Κρήτης (in Markomichelaki 2014, 90).

22 On the icon, see Vocotopoulos (1990, 99–100, 1988, 223); Chondrogiannis (2010, 46).

23 Cf. Cormack (1985, 50–94); Papamastorakis (1998, 222). On the miracles of St. Demetrios, see Lemerle (1979–1981); Bakirtzis (1997).

24 Bakirtzis (2012, 167–169, Fig. 47); Cormack (1985, 51–54, Fig. 14).

25 Xyngopoulos (1936, 108, Pl. B.5, Fig. 5); Grabar (1950, 3–5); Christou, Mavropoulou-Tsioumi, Kadas, and Kalamartzi-Katsarou (1991, Fig. on 21).
26 Metcalf (1966, 259, Fig. on frontispiece); Papamastorakis (1998, 227, Fig. 11).
27 Radovanović (1987, 82–85, Figs 9–11).
28 Bakirtzis (2002, 187–192).
29 Mazower (2006, 119).
30 'Ω μεγαλομάρτυρα του δεσπότου Χριστού Δημήτριε, που είναι τα θαύματα, οπού καθ' εκάστην ημέραν έκαμνες εδώ εις την πατρίδα σου, πώς τώρα δεν μας βοηθείς; Πώς δεν μας επισκέπτεσαι; Διατί, άγιε Δημήτριε, αστόχησες και τελείως μας απάργιασες; Δεν βλέπεις πόσα βάσανα μας επερικύκλωσαν, πόσοι πειρασμοί, πόσα χρέη; Δεν βλέπεις πώς εγεννήθημεν όνειδος και καταπάτημα των εχθρών, παίγνιο των ασεβών, περίπαιγμα των Σαρακηνών, εξουθένημα και γέλως των πάντων; Διατί λοιπόν, άγιε, δεν μας λυπάσαι ως συμπατριώτης; Διατί δεν μας συμπονάς; Δεν ακούεις τους αναστεναγμούς, δεν βλέπεις τα δάκρυά μας; Δεν στοχάζεσαι τους παραδαρμούς μας;' The enkomion, read on the feast day of the saint in a church dedicated to him, in Thessaloniki, near the Church of Hagios Gregorios Palamas, is included in cod. 76, at the Stavroniketa Monastery, Mount Athos. Laourdas (1955–1960, 113–115, 156–159); Mazower (2006, 119).
31 Tsiaples (2016, 77–80); Gouloulis (2015, 250–253); Kaltsogianni, Kotsabassi, and Paraskevopoulou (2002, 116, n. 163); Tsiaples (2014, 143–145).
32 'Πού μοι, Δημήτριε μάρτυς, η αήττητος συμμαχία; Πως την σην πόλιν υπερείδες πορθούμενην; Πως υπό σοι πολιούχω η εχθροίς άβατος, αφ' ου χρόνου ταύτην, ήλιος εθεάσατο, τοσούτων κακών εις πείραν εγένετο; Πως της των δυσσεβούντων οφρύος ηνέσχου κατορχουμένων της ιεράς προστασίας; Πως υπέμεινας ταύτα και διεκαρτέρησας' (Westerink 1981, 10–13).
33 On the icon, see Albani (2016).
34 The fire, which destroyed the birthplace of Bosch, 's-Hertogenbosch (or Den Bosch), in 1463, remained alive in his memory, and so he favoured the motif of burnt down cities.
35 Buendía, Cruz Valdovinos, Gutiérrez Pastor, Morales Folguera, and Rincòn García (1994, 376, Pls on 378, 380–381).
36 Slatkes (1975); Hills (2007, 187).
37 Gibson (1992, 212, Fig. 14). The panel illustrates a medieval text (*Visio Tnugdali*) by Brother Marcus, an Irish itinerant Benedictine monk, reporting the otherworldly vision of the Irish knight Tnugdalus. It was written in Latin shortly after 1148 and translated into 15 languages by the 15th century. Easting (1997, 70–80); Spilling (1975).

Bibliography

Aalberts, J. 1998. 'He Hierousalem kai he Babylona. He Halose tes Poles kai he Synteleia tou Kosmou (Jerusalem and Babylon. The Fall of the City and the End of the World).' In *Kretikes Spoudes (Cretan Studies)* 6, 241–265.

Albani, J. 2016. 'The Son of Man is Coming' (Matthew 24:44). An Icon of the Last Judgment from the Church of Virgin Faneromenei, Lefkosia.' In Dželebdžić, D. and Bojanin, S. eds., *Proceedings of the 23rd International Congress of Byzantine Studies*. Belgrade, 22–24.08.2016. Thematic Sessions of Free Communications. Belgrade [n. pub], 467–468.

Albani, J. 2017. 'Between Babylon and Holy Zion: Remarks on a 16th-century Icon with the Allegory of Heavenly Jerusalem.' In Constantoudaki-Kitromilides, M. ed., *Painting and Society in Venetian Crete. Evidence from Portable Icons. An International Symposium*. Athens [n. pub.], 20.

Albani, J. Forthcoming. 'Between Babylon and Holy Zion: Remarks on a 16th-century Icon with the Allegory of Heavenly Jerusalem.' In Buschhausen, H. and Prolović, J.

eds., *Erforschen – Erkennen – Weitergeben. Gewidmet dem Gedenken an Helmut Buschhausen*. Vienna. Belgrade: Publisher to be announced.

Bakirtzis, Ch. 1997. *Hagiou Demetriou Thaumata. Hoi Sylloges Archiepiskopou Ioannou kai Anonymou. Ho Bios, ta Thaumata kai he Thessaloniki tou Hagiou Demetriou (Miracles of St. Demetrios. The Collections of Archbishop John and of Anonymous)*. Introduction, comments and edition by Bakirtzis, Ch., translation by Sideri, A. Thessaloniki: Agra.

Bakirtzis, Ch. 2002. 'Pilgrimage to Thessaloniki: The Tomb of St. Demetrios.' In *Dumbarton Oaks Papers* 56, 175–192.

Bakirtzis, Ch. 2012. 'Hagios Demetrios (St. Demetrios).' In Bakirtzis, Ch. ed., *Psephidota tes Thessalonikis, 4os – 14os Aionas (Mosaics of Thessaloniki, 4th – 14th Century)*. Athens: Ekdoseis Kapon, 128–179.

Baskins, C., Randolph, A.W.B., and Musaccjio, J.M. 2008. *The Triumph of Marriage. Painted Cassoni of the Renaissance*. Boston: Isabella Gardner Museum in Association with Gutenberg Periscope Publishing.

Buendía J.R., Cruz Valdovinos J.M., Gutiérrez Pastor I., Morales Folguera J.M., and Rincòn García W. 1994. *Paintings of The Prado*. Boston, New York and London: Little, Brown and Company.

Campbell, C.J. 2001. 'The City's New Clothes: Ambrogio Lorenzetti and the Poetics of Peace.' In *The Art Bulletin* 83(2), 240–258.

Chondrogiannis, S. Th. 2010. *The Antivouniotissa Museum, Corfu*. Thessaloniki: M. Diamantidi.

Chondroyannis, S. 2010. 'Icon of the Journey to the New Jerusalem, the Ascent from Earth to the Heavenly City.' In Ćurčić, S. and Hadjitryphonos, E. eds., *Architecture as Icon. Perception and Representation of Architecture in Byzantine Art*. Princeton, NJ: Princeton University Art Museum, 340–347.

Christou P.K., Mavropoulou-Tsioumi Ch., Kadas S.N., and Kalamartzi-Katsarou Ai. 1991. *Hoi Thesauroi tou Hagiou Orous*. Seira A´. *Eikonographemena Cheirographa, Parastaseis – Epititla – Archika Grammata*. Tomos, D., M. Batopediou, M. Zographou, M. Stauroniketa, M. Xenophontos (*The Treasures of the Holy Mountain*. Series 1. *Illuminated Manuscripts, Representations – Headings – Initials*, vol. 4. *Vatopedi Monastery, Zographou Monastery, Stavroniketa Monastery, Xenophontos Monastery*). Athens: Ekdotike Athenon.

Cormack, R. 1985. *Writing in Gold. Byzantine Society and its Icons*. London: George Philip.

Dupont, J. and Gnudi, C. 1990. *La Peinture Gothique*. Geneva: Skira.

Easting, R. 1997. *Annotated Bibliographies of Old and Middle English Literature III. Visions of the Other World in Middle English*. Cambridge: D.S. Brewer.

Eisenbichler, K. and Iannucci, A.A. 1990. *Petrarch's 'Triumphs': Allegory and Spectacle*. Ottawa: Dovehouse Editions.

Finotti, F. 2009. 'The Poem of Memory (*Triumphi*).' In Kirkham, V. and Maggi, A. eds., *Petrarch. A Critical Guide to the Complete Works*. Chicago and London: The University of Chicago Press, 63–83.

Georgopoulou, M. 1995. 'Late Medieval Crete and Venice: An Appropriation of Byzantine Heritage.' In *The Art Bulletin* 77(3), 479–496.

Gibson, W.S. 1992. 'Bosch's Dreams: A Response to the Art of Bosch in the 16th Century.' In *The Art Bulletin* 74(2), 205–218.

Gouloulis, S.G. 2015. 'Ho Hagios Demetrios hos Stratiotes ektos Thessaloniki. Dyo Emphaniseis tou sten Helladike Hodo ton Proskyneton (Diegeseis B 6 kai C 3) (St. Demetrios as a Soldier out of Thessaloniki. His two Visionary Appearances in the Greek Road of Pilgrims (Narrations II 6 and III 3]).' In Katsaros, B. and Tourta, A. eds., *Aphieroma* ston Akademaiko Panagiote L. Bokotopoulo (Festschrift for Academician Panagiotis L. Vocotopoulos). Athens: Ekdoseis Kapon, 247–256.

Grabar, A. 1950. 'Quelques Reliquaires de St. Démétrios et le Martyrium du Saint à Salonique.' In Dumbarton Oaks Papers 5, 2–28.

Gratziou, O. 2016. 'From Heaven to Earth. Perceptions of Reality in Icon Painting.' In Ikon 9, 53–64.

Hills, P. 2007. 'Titian's Fire: Pyrotechnics and Representations in 16th-Century Venice.' In Oxford Art Journal 30(2), 187–204.

Kaltsogianni, E., Kotzabassi, S. and Paraskevopoulou, I. 2002. He Thessaloniki ste Byzantine Logotechnia. Retorika kai Hagiologika Keimena (Thessaloniki in Byzantine Literature. Rhetorical and Hagiological Texts). Thessaloniki: Byzantine Research Centre.

Kephallonitou, F. 1994. 'Allegory of Heavenly Jerusalem.' In Albani, J. ed., Icons Itinerant. Corfu, 14th – 16th Centuries. Athens: Ministry of Culture, 150–155.

Laourdas, B. 1955–1960. 'Byzantina kai Metabyzantina Egkomia eis ton Agion Demetrion (Byzantine and Post-Byzantine Enkomia in Honour of St. Demetrios).' In Makedonika 4, 47–162.

Lemerle, P. 1979–1981. Les plus Anciens Recueils des Miracles de St. Démétrius et la Pénétration des Slaves dans les Balkans, 2 vols. Paris: Centre National de la Recherche Scientifique.

Lidov, A. 1998. 'Heavenly Jerusalem: The Byzantine Approach.' In Jewish Art, 340–353.

Maguire, H. 1981. Art and Eloquence in Byzantium. Princeton: Princeton University Press.

Maguire, H. 2012. Nectar and Illusion. Nature in Byzantine Art and Literature. Oxford and New York: Oxford University Press.

Mango, C. 1958. The Homelies of Photius Patriarch of Constantinople. English Translation, Introduction and Commentary. Cambridge, MA: Harvard University Press.

Markomichelaki, T.M. 2014. Manoles Sklabos. Tes Kretes ho Chalasmos (He Symphora tes Kretes) (Manuel Sklavos. The Destruction of Crete [The Calamity of Crete]). Thessaloniki: Aristoteleio Panepistemio Thessaloniki. Institouto Neoellenikon Spoudon (Hidryma Manole Triantaphyllide).

Markomichelaki, T.M. 2015. Edo, eis to Kastron tes Kretes . . . Henas Logotechnikos Chartes tou Benetsianikou Chandaka (Here, at the Kastron of Crete . . . A literary Map of Venetian Chandax). Thessaloniki: University Studio Press.

Mastora, P. 2012. 'Peri tes Ano Hierousalem (On Heavenly Jerusalem).' In Eikonostasion 3, 47–60.

Mazower, M. 2006. Thessaloniki. Pole ton Phantasmaton. Christianoi, Mousoulmanoi kai Hebraioi, 1430–1950 (Thessaloniki. City of Ghosts. Christians, Muslims and Jews, 1430–1950), translation by Kouremenos, K. Athens: Alexandreia.

Metcalf, D.M. 1966. Coinage in the Balkans, 820–1355. Chicago: Argonaut Publishers.

Muir, E. 1979. 'Images of Power: Art and Pageantry in Renaissance Venice.' In The American Historical Review 84(1), 16–52.

Papamastorakis, T. 1998. 'Histories kai Historeseis Byzantinon Pallekarion (Tales and Images of Byzantium's Warrior Heroes).' In Deltion of the Christian Archaeological Society, 4th ser., 20, 213–230.

Polzer, J. 2002. 'Ambrogio Lorenzetti's "War and Peace" Murals Revisited: Contributions to the Meaning of the "Good Government Allegory".' In Artibus et Historiae 23(45), 63–105.

Pope-Hennesy, J. and Christiansen, K. 1980. 'Birth Trays, Cassone Panels, and Portraits.' In The Metropolitan Museum of Art Bulletin, Summer, 14–63.

Radovanović, J. 1987. 'Heiliger Demetrius – Die Ikonographie seines Lebens auf den Fresken des Klosters Dečani.' In Samardžić, R. and Davidov, D. eds., L'Art de

Thessalonique et des Pays Balkaniques et les Courants Spirituels au XIVᵉ Siècle. Recueil des Rapports du IVᵉ Colloque Serbo-grec, Belgrade 1985. Belgrade [n. pub.], 75–88.

Skinner, Q. 1999. 'Ambrogio Lorenzetti's Buon Governo Frescoes: Two Old Questions, Two New Answers.' In *Journal of the Warburg and Courtauld Institutes* 62, 1–28.

Slatkes, L.J. 1975. 'Hieronymus Bosch and Italy.' In *The Art Bulletin* 57(3), 335–345.

Spilling, H. 1975. *Die Visio Tnugdali. Eigenart und Stellung in der Mittelalterlichen Visionsliteratur bis zum Ende des 12. Jahrhunderts*. Munich: Arbeo-Gesellschaft.

Triantaphyllopoulos, D.D. 2016. 'Eschatologia, Aretologia kai Techne. Noemata Mias Eikonas sten Kerkyra sta Tele tou 15ou Aiona (Eschatology, Aretology and Art. Meanings of an Icon in Corfu Dated around the End of the 15th Century).' In *Kerkyraika Chronika* 9, 593–604.

Tsiaples, G.V. 2014. *Poliorkies kai Haloseis Poleon sta Byzantina Retorika kai Hagiologika Keimena (Sieges and Captures of Cities in Byzantine Rhetorical and Hagiological Texts)*. Unpublished doctoral thesis. Thessaloniki: Aristoteleio Panepistemio Thessalonikis. Available at: http://hdl.handle.net/10442/hedi/35295 (Accessed 10.04.2019).

Tsiaples, G.V. 2016. 'Mythiko Parelthon – Christianiko Paron: Proslepse kai Probole ton Pneumatikon Scheseon dyo Byzantinon Poleon. To Paradeigma tes Thessalonikis kai tes Larisas (9os–14os ai.) [Mythical Past – Christian Present: Perception and Fostering of Spiritual Ties between two Byzantine Cities. The Case of Thessaloniki and Larisa (9th–14th century)].' In *Byzantina Symmeikta* 26, 67–92.

Tsimpoukis, G.D. 2013. *He Apokalypse tou Ioanne ste Mnemeiake Zographike tou Hagiou Orous (The Revelation of John in Monumental Painting at the Holy Mountain)*. Athens: Bookstars Editions.

Vocotopoulos, P.L. 1988. 'St. Demetrius.' In Acheimastou-Potamianou, M. ed., *Holy Image, Holy Space. Icons and Frescoes from Greece*. Athens: Greek Ministry of Culture. Byzantine Museum of Athens, 223.

Vocotopoulos, P.L. 1990. *Eikones tes Kerkyras (Icons of Corfu)*. Athens: Morphotiko Hidryma tes Ethnikes Trapezes.

Vocotopoulos, P.L. 2016. 'Renaissance Influence on Post Byzantine Panel Painting in Crete.' In Tuchkov, I. et al. eds., *Actual Problems of Theory and History of Art. VI. Collection of Articles*. St. Petersburg and Moscow: The State Hermitage Museum and Kremlin Museums, 177–184.

Westerink, L.G. 1981. *Nicholas I Patriarch of Constantinople. Miscellaneous Writings. Greek Text and English Translation*. Washington, DC: Dumbarton Oaks.

White, J. 1993. *Art and Architecture in Italy, 1250–1400*. New Haven and London: Yale University Press, Pelican History of Art.

Wilkins, E.H. 1963. 'The First two Triumphs of Petrarch.' In *Italica* 40(1), 7–17.

Wyss, E. 1998. 'A "Triumph of Love" by Frans Francken the Younger: From Allegory to Narrative.' In *Artibus et Historiae* 19(38), 43–60.

Xyngopoulos, A. 1936. 'Byzantinon Kibotidion meta Parastaseon ek tou Biou tou Hagiou Demetriou (Byzantine Box with Scenes from the Life of St. Demetrios).' In *Archaiologike Efemeris*, 101–136.

Part 3

Current crisis and urban insurgency as contestation of the urban sublime

From generic theory to Athenian praxis

Map 4 Central Athens and neighbouring municipalities. The Exarcheia neighbourhood is the central triangle-like area.

Source: © Wikipedia Commons. Source: https://el.wikipedia.org/wiki/Εξάρχεια#/media/Αρχείο:Location_map_Greece_Athens_central.png

7 The political art of urban insurgency[1]

Erik Swyngedouw

Art, politics, truth

Alain Badiou insists that there are four registers of truth that constitute a possible truth-event. In such truth-event, subjects that might inaugurate a truth procedure begin to emerge. These four registers of truth are: Art, Love, Science, and Politics (Badiou 2005). A truth procedure is the process through which an act or intervention – that is an active bodily performance – becomes retroactively inaugurated as a truth-event; such an event inaugurates a rupture in the Heideggerian order of being, in the state-of-the-situation. It transforms the co-ordinates of being in a given condition and opens potentially a process of transformation of the politico-institutional order. Such an event may open up a truth procedure through the declaration of fidelity to such inaugural moment. The truth-procedures associated with each of the four registers of truth (art, love, science, politics) differ fundamentally in terms of the count that declares the truth of the event. The truth of Love and of Science operates under the count of Two. The truth of Love resides in the affirmation of its truth by the one to whom it is declared and who assumes fidelity to the truth-event: the event of Love. When I say 'I love you' and the person to whom this declaration is made responds with 'I love you too,' a fidelity to the truth of this event will set in motion a procedure with life-shattering consequences for the two subjects involved. The truth of a scientific statement resides in the affirmation of its truth by another scientist, who bears witness to the truth of the scientific event. The truth of Art is singular; it resides in the declaration of the truth of Art by the One proclaiming its truth and declaring his or her fidelity to the truth of the event: 'I declare this work to be a work of Art.' Examples here are of course Duchamp's urinal or Jeff Koons in pornographic performance with his then wife, the porn star *La Cicciolina*. Badiou (2003) claims that 'today art can only be made from the starting point of that which, as far as Empire is concerned, doesn't exist. Through its abstraction, art renders this inexistence visible. This is what governs the formal principle of every art: the effort to render visible to everyone that which for Empire (and so by extension for everyone, though from a different point of view), doesn't exist.'

The truth of politics, of course, resides in the multiple, the many, the militant fidelity of a universalizing multitude to an inaugural emancipatory event (like

October 1917, The Paris Commune, or May '68). It is the affirmation of the truth perceptible in an inaugural political event and unfolds through the universalization of declarations of fidelity to the event, an interruption that retroactively can be designated as a political event (Badiou 2006). A political event is a collectively staged interruption in the state-of-affairs (consider Syntagma Square in 2011 and beyond), which may universalise or spatialise through the declaration of fidelity of others to the meaning inscribed in the event. Such universalization is precisely the process of politicisation, announcing the potential transformation of the state-order (for example, the surprise election of Syriza in 2015 and its first six months of heroic anti-austerity politics).

The relationship between art and politics, therefore, resides fundamentally in the articulation between two distinct truth procedures, establishing the relationship between the one and the multiple, the singular and universal. Let me illustrate this reasoning. When on December 1, 1955 in Montgomery, Alabama, African American civil rights activist Rosa Parks sat down on a whites-only bus seat to go home after work, Rosa could have declared it a work of Art, a gesture that would be reminiscent of Duchamp's declaration of the urinal as a work of Art or of the bodily performances of Marina Abramović. In contrast, Rosa's intervention, while individual, was aimed at the universal, performing equality and exposing the 'wrong' of the existing politico-legal and institutional order (what Jacques Rancière designates as 'the police'). Although the US Constitution declared and affirmed the axiomatic equality of each and every one qua speaking and thus political beings, the actually existing situation – the police – insisted on sustaining a radical inegalitarian and racialized social order. As had happened many times before, Rosa's act was indeed properly policed by the bio-political power of the state. She was arrested. However, the retroactive declaration of Rosa's minimal intervention in the common-sense order of the day as a political event – one of the symbolic inaugural events of what later would be symbolized as the civil rights movement and that the National Association for the Advancement of Colored People (NAACP) had planned carefully – resided in the subsequent declaration of a multitude, with highly varied gender, ethnic, racial, or class inscriptions, of their fidelity to the truth of the emancipatory act performed by Rosa's minimal disturbance (May 2008). Many heterogeneous people joined up, declaring 'I am Rosa Parks too.' An emancipatory politicizing sequence unfolded and intensified subsequently, one that would turn out to be highly performative, under the name of the Civil Rights Movement. This politicizing sequence would transform the co-ordinates of collective life and produce a new common sense that radically replaced and declared as utterly nonsensical the common sense that prevailed before the act (i.e., that it was perfectly sensible to exclude coloured or black people from certain urban places and functions).

Or consider the following, more recent, example of the articulation between art and urban insurgency. On November 3, 2014, just a few days before the commemoration of the 25th anniversary of the fall of the Berlin Wall (an unparalleled urban artistic political intervention in itself), a group of activist artists removed seven crucifixes inscribed with the names of the East Berliners that had died in their fateful attempt to cross the divide between East and West during the time that Berlin was a divided city. The crucifixes were located next to the German *Reichstag* building on the banks of the River Spree to bear witness to the perverse effects and intolerable suffering of walls and their separating and dividing functions. In their stead, a sign with the inscription '*There is no thinking going on here*' was left. This event was followed immediately by a major political and media attack condemning this act of brutal vandalism as defiling the wounds and pains of a once divided community.

However, after a few days, the seven crosses re-appeared on the outer steel and concrete borders of the European Union in Bulgaria, Greece, and the Spanish enclaves in Northern Africa (Melilla and Cueta) where they were carried by refugees who tried to enter the European Union but were kept outside behind barbed wire. Through this intervention, the artists re-inscribed the memory of the divided and walled city in the actually existing geo-exclusive politics and socially triaging practices of a gated Europe. Here too, the *Zentrum für Politische Schönheit* (ZPS; Centre for Political Beauty) as the artists call themselves, aspired to a de-aesthetization of contemporary post-democratic politics and its exclusionary common sense by means of re-articulating the semantic enchainment of Walls, Division, Separation, and Suffering to a new common sense, a new distribution of the perceptible, around the universalizing signifiers of equality, freedom, and inclusive humanity.[2] In the aftermath, the ZPS was investigated for 'forming a criminal organization.' The investigation was closed in April 2019 without charges.

These vignettes illustrate how both Art and Politics dwell in the register of the Aesthetic understood in a Rancièrian mode as 'the distribution of the sensible,' the partitioning of what can be heard, voiced, sensed, felt, registered, and through which a common sense becomes configured. Artistic and political interventions are interruptions in the senses – in the aesthetic register – with a view to make sensible what is non-sense, what is only registered as noise. Such acts signal in their performative staging the inegalitarian forms of the existing state of the situation, and, through their interruptional performativity, address 'the wrong' of the inegalitarian condition while performing, voicing, rendering sensible, what 'equality' is substantively all about. While the truth of Art remains singular, the truth of Politics is a universalizing one.

Indeed, as Jacques Rancière keeps reminding us, Art and Politics revolve around similar aesthetic procedures. We are not referring here to the aestheticization of politics that Walter Benjamin identified with Fascism, or that can also be traced in contemporary forms of consensual and aestheticized post-political forms of technocratic management of the givens of the situation (see Swyngedouw 2018), but rather to politics (and art) as an aesthetic procedure. Here, aesthetics is not understood in its 19th-century rendition of that what is related to 'beauty,' but in a classical sense as to what is accessible to the senses. Both aesthetics and the senses revolve around the registers of the perceptible. As Rancière notes:

> In the end everything in politics turns on the distribution of spaces. What are these places? How do they function? Why are they there? Who can occupy

them? For me, political action always acts upon the social as the litigious distribution of places and roles. It is always a matter of knowing who is qualified to say what a particular place is and what is done to it.

(Rancière 2003, 201)

For Jacques Rancière, artistic practices, like political interruptions, are an integral part of the partition of the perceptible to the extent that they suspend the common sense of sensory experience and reframe the relationships between subjects and objects, the common and the singular, what is visible and what is not, one place and another. With Alain Badiou's notion of the event and Jacques Rancière's conceptualization of art and politics as interruptions in the order of the sensible – a disruption of the police order – we may begin to discern how urban art and emancipatory urban politics intertwine. Indeed, the articulations between Art and Politics find their expression precisely in the spatial, in the urban. As David Harvey keeps insisting, every political project is a spatial one, and every spatial project is a political one (Harvey 2012). And the notion of 'the political' has to be taken in the precise meaning that Alain Badiou, Jacques Rancière, or Chantal Mouffe, albeit in different ways, assign to 'the political' as an immanent moment of interruption, an intervention that destabilizes the order of being and aspires to an infinitely inclusive universalization, one that destabilizes 'politics' as the existing common-sense modalities of being-in-common (Mouffe 2005; Wilson and Swyngedouw 2014).

Such 'political' interventions stand indeed in stark contrast to the celebration of the relationship between (public) art and politics understood as the everyday choreography of public management. The radical difference between an urban political intervention and an intervention that re-enforces the suffocating and often violent hold that everyday politics/the police exerts over the potential emergence of a political event can be easily exemplified by two actions that took place in Amsterdam a few years ago. In an act of urban guerrilla intervention, American-Italian artist Arturo DiModica placed his third (after New York in 1989 and Shanghai in 2010) Charging Bull (of Wall Street fame) on Amsterdam's Beursplein (Exchange Square) on July 4, 2012 (not co-incidentally US Independence Day), just a little while after the site was cleared off a small coterie of Occupy! activists, too small in numbers to even itch the powers that be or attract international attention. Is the Charging Bull's presence – the triumphant symbol of a victorious capitalo-financial-parliamentary order – not one of the most tell-tale signs of the symbolic re-appropriation of urban space by the 1%, by those who resist change, who defend the prevailing common sense by all means available? While the square was cleared of its protesters, the unauthorized intrusion of the Bull was quickly legitimized and approved by the city administrators. It now takes pride of place as Amsterdam's elite made quite clear to all what the Beursplein and city politics stand for; an urbanity manicured to nurture the circulation of money and capital accumulation. It signals, in this case, the mobilization of the aesthetics of guerrilla urban interventions to ensure that nothing really changes, to re-affirm that Art and Politics can seamlessly fuse together to sustain the existing while suppressing the nurturing of urban interventions that seek to transform the state of the situation, that aim to re-order the common sense of everyday life, one that embryonically signals the desire for the inauguration of a new politics.

It is from this perspective that I wish to explore further in this contribution the *Political Art of Urban Insurgency*, as manifested in the seemingly never ending proliferation of urban rebellions, unfolding against the backdrop of very different historical and geographical contexts, that – since the magically riotous year of 2011 – profoundly disturbed the apparently cosy neoliberal status-quo and disquieted various economic and political elites. There is indeed an uncanny choreographic affinity between the eruptions of discontent in cities as diverse as Istanbul, Cairo, Tunis, Athens, Madrid, Hong Kong, New York, Cape Town, Tel Aviv, Chicago, London, Berlin, Thessaloniki, Santiago, Stockholm, Barcelona, Montreal, Oakland, São Paulo, or Paris, among many others. A wave of profoundly urban interruptions is rolling through the world's cities, whereby those who do not count, 'the part of no-part,' demand a new constituent process for producing space politically. Under the generic name of '*Real Democracy Now!*,' the heterogeneous mix of gatherers are outraged by and expose the variegated 'wrongs' and spiralling inequalities of autocratic neo-liberalization and actually existing instituted democratic governance. A politicized and disruptive mobilization, animated by an eclectic mix of insurgent urban architects, is increasingly choreographing the contemporary theatre of urban politicized struggle and conflict (Swyngedouw 2013). From a radical urban political perspective, the central question that has opened up, after the wave of insurgencies of the past few years petered out, revolves centrally around what to do and what to think next. Is there further thought and practice possible after the squares are cleared, the tents broken up, the art destroyed, the energies dissipated, and everyday life resumes its routine practices? What can be learned from those ephemeral experiences that nonetheless nurture the continuous return of *Spaces of Hope* (Harvey 2000)?

The spectral return of the political

For Jacques Rancière, democratizing the Polis is inaugurated when those who do not count stage the count, perform the process of being counted, and thereby initiate a rupture in the order of things, 'in the distribution of the sensible,' such that things cannot go on as before (Rancière 1998). From this perspective, democratization is a performative act that both stages and defines equality, exposes a *wrong*, and aspires to a transformation of the senses and of the sensible, to render common sense of what was non-sensible (and considered to be non-sense) before. This is where Art and Politics as aesthetic procedures meet. Democratization, Rancière contends, is a disruptive affair whereby the *ochlos* (the rabble, the crowd, the scum, the outcasts, 'the part of no part') stages to become part of the *demos* (a citizen of the Polis) and, in doing so, inaugurates a new ordering of times and places, a process by which those who do not count, who do not exist as part of the Polis become visible, sensible, and audible, stage the count, and assert their egalitarian existence. Egalitarian politics is about 'the symbolic institution of the political in the form of the power of those who are not entitled to exercize power – a rupture in the order of legitimacy and domination. It is the paradoxical power of those who do not count: the count of the "unaccounted for" ' (Rancière 2000). Egalitarian-democratic demands and practices, scandalous in the representational order of the police yet eminently realizable, are precisely those staged through mobilizations varying from the Paris and Shanghai communes to the Occupy!, *Indignado!*, and assorted other emerging urban political movements that express and nurture such processes of embryonic re-politicization (Dikeç and Swyngedouw 2017). Occasionally such interventions are spectacular, often they are mundane as in the case of the performative re-ordering of the spaces where state sovereignty is suspended, such as in the use of Churches, for example, in Brussels, Paris, or London, when undocumented migrants occupy such sacred spaces and turn them into sites of asylum and places for the performance of equality. Identitarian positions become, in the process, transfigured into a commonality, and a new common sense – it is the political at work through the process of political subjectivation, of acting in common by those who do not count, who are excessive, surplus to the police.

There are many uncounted today. Alain Badiou refers to them as the 'inexistent,' the masses of the people that have no say, 'decide absolutely nothing, have only a fictional voice in the matter of the decisions that decide their fate' (Badiou 2012, 56). These inexistent are the motley assortment of a-political consumers, frustrated democrats, precarious workers, insurgent architects and artists, undocumented migrants, recalcitrant intellectuals, and disenfranchized citizens. The scandal of actually existing, instituted (post-)democracy in a world choreographed by oppression, exploitation, and extraordinary inequalities resides precisely in rendering masses of people inexistent, politically mute, without a recognized voice.

For Alain Badiou, 'a change of world is real when an inexistent of the world starts to exist in the same world with maximum intensity' (Badiou 2012, 56). In doing so, the order of the sensible is shaken and the kernel for a new common sense, a new mode of being in common becomes present in the world, makes its presence sensible and perceptible. It is the appearance of another world in the world, and as such a profoundly aesthetic affair. What was considered normal, right, and beautiful cannot any longer be perceived or lived as such; new registers of senses, voices, and semantic sequences emerge, something that Antonio Gramsci would consider to announce a shift in hegemony, produced through the mobilizing force of universalizing, and thus spatializing, counter-discourses and counter-practices. Has it not been precisely the sprawling urban insurgencies since 2011 that ignited a new sensibility about the Polis as a democratic and potentially democratizing space? This appearance of the inexistent, staging the count of the uncounted is, it seems to me, what the Polis, the political city, is all about. Indeed, as Foucault reminds us, '[t]he people is those who, refusing to be the population, disrupt the system' (Foucault 2007, 43–44).

The notion of the democratizing Polis introduced previously is one that foregrounds intervention and rupture, and destabilizes the apparently cosy bio-political order, sustained by an axiomatic presupposition of equality. Democratization, then, is the act of the few who become the material and metaphorical stand-in for the many; they stand for the dictatorship of the democratic – direct and egalitarian – against the despotism of the instituted 'democracy' of the elites – representative and inegalitarian (Badiou 2012, p. 59). Is it not precisely these insurgent architects that brought to the fore the irreducible distance between the democratic as the immanence of the presupposition of equality on the one hand and democracy as an instituted form of regimented oligarchic techno-managerial governing on the other? Do the urban revolts of the past few years not foreground the abyss between 'the democratic' and 'democracy,' the surplus and excess that escapes the suturing and de-politicizing practices of instituted governing? Is it not the re-emergence of the proto-political in the urban revolts that signals an urgent need to re-affirm the urban, the polis, as a political space, and not just as a space of bio-politically governed city life?

Of course, the social markers of the insurgencies are geographically highly differentiated. The quilting points around which the interruption becomes articulated are invariably particular, specific, and concrete: a threatened park and a few trees in Istanbul making place for the vernacular architecture of an Ottoman mosque and shopping mall, an authoritarian regime in Egypt, massive austerity in Greece, Portugal and Spain, social and financial mayhem in the UK or the US, a rise in the price of public transport tickets in São Paulo, the further commodification of higher education in Montreal, large-scale gold-mining in Rumania. Yet, the urban insurgents quickly turned their particular, occasionally identitarian, grievances into a wholesale attack on the instituted order, on the unbridled commodification of urban life in the interests of the few, on the highly unequal socio-economic outcomes of actually existing representational *post-democracy-cum-capitalism*. The particular demands transformed quickly and seamlessly into a universalizing staging for something different, however diffuse and unarticulated this may presently be. The assembled groups ended up without particular demands addressed to the elites, to a Master. In their refusal to express specific grievances, they demanded everything, nothing less than the transformation of the instituted order.

In their urban socio-spatial interruptive acting, they staged new ways of practicing equality and democracy, experimented with innovative and creative ways of being together in the city, and prefigured, both in practice and in theory, new ways of distributing goods, accessing services, producing healthy environments, organizing debate, managing conflict, practicing ecologically saner life-styles, and negotiating urban space in an emancipatory manner. The dominant aesthetic and semantic registers of common sense were blown apart while others embryonically emerged and were experimented with.

These insurgencies are decidedly urban; they may be the manifestation of the immanence of a new urban commons (García Lamarca 2013), one always potentially in the making, aspiring to produce a new urbanity through intense meetings and encounters of a multitude, one that aspires to spatialization, that is to universalization. Such universalization can never be totalizing as the demarcation lines are clearly drawn, lines that separate the 'us' (as emancipatory multitude) from the 'them', that is, those who mobilize all they can to make sure nothing really changes.

The democratizing minority stands here in strict opposition to the majoritarian rule of instituted democracy. As much as the proletarian, feminist or African American democratizing movements were (and often still are) also very much minoritarian in terms of politically acting subjects, they nonetheless stood and stand for the enactment of the democratic presumption of equality of each and all. The space of the political disturbs the socio-spatial ordering by re-arranging it with those who stand in for 'the people' or 'the community' (Rancière 2001). It is a particular that stands for the whole of the community and aspires toward universalization. The rebels on Tahrir, Syntagma, or Taksim Square are not the Egyptian, Greek, or Turkish population; while being a sociological minority, they stand materially and metaphorically for the Egyptian, Greek, and Turkish People. The political emerges, Rancière attests, when the few claim the name of the many, to embody the community as a whole, and are recognized, sensed, as such. The emergence of political space is always specific, concrete, particular and minoritarian, but stands as the metaphorical condensation of the generic, the many, and the universal.

These attempts to produce a new commons offer perhaps a glimpse of the theoretical and practical agenda ahead. Does their acting not signal a clarion call to return the intellectual gaze to consider again what the polis has always been, namely the site for political encounter and place for enacting the new, the improbable, things often considered impossible by those who do not wish to see any change, the site for experimentation with, the staging and production of new radical imaginaries for what urban democratic being-in-common might be all about?

The artistic violence of urban insurrection

Indeed, the ultimate aim of urban interventions is to change the given socio-spatial ordering in a certain manner. Like any intervention, this is a violent act. It erases at least partly what is there in order to erect something new and different. Interruption is precisely predicated upon a voluntarist act of subjective violence, the will to place one's body, voice, and practice in the circuits and flows of the given, and to expose oneself, to render oneself vulnerable to the violence of those who insist that life has to go on, that the flow and the circuits require uninhibited circulation and movement. In the face of attack or interruption, the key thing to do, the elites tells us, as George Bush, Jr., allegedly reminded us after 9/11, is 'to keep on shopping.'

It is of central importance to recognize that politicizing acts are singular interventions that (aspire to) produce particular socio-spatial arrangements and urban milieus and, in doing so, foreclose (at least temporarily) the possibility of others to emerge. Any intervention enables or foreshadows the formation of certain socio-spatial matrices and closes down others. The subjective 'violence' inscribed in such choice has to be fully endorsed and its implications teased out.

The violence of urban insurrection and interruption is subjective; the result of a voluntarist decision to act, to stage the new, the different and, in the process, make visible the objective violence inscribed in, and often actively mobilized by, the instituted bio-political police order. While objective violence is precisely embodied and performative through the common-sense operation of the existing order that renders some more equal than others, that relegates some to the margins of life, that reduces some bodies to bare life, that triages functions and places such that parts are distributed differentially, that some fall out of the count, become 'the part of no part.' Subjective violence, therefore, is an aesthetic practice, a collective work of art, of rendering visible, audible, and articulate that what the objective violence of the police order renders mute, invisible, inaudible, and inarticulate. Here too resides the impassable abyss between subjective violence of identitarian, and therefore exclusive, interruptions (like, for example, the unspeakably horrible brutal murdering of people at Bataclan in Paris or in Brussels's subway) and the subjective violence of democratizing and emancipatory urban interruptions. An egalitarian politics is radically inclusive; everybody is invited in, 'it is an inclusionary struggle' (Žižek 2013) that cuts through the socially segregating borders and boundaries of race, class, ethnicity, gender, location, or sub-alterity. Of course, the question then arises of how to confront those who remain on the outside, who will mobilize whatever dispositive to prevent the universalization of the inclusionary struggle. Against their symbolic and objective violence, it is vital to think about ways to protect and defend the universalizing process without descending into abyssal terror, about how to navigate the prospect of failure in the absence of effective defence as experienced by the Paris Commune or in the violence of political terror that marked so much of past emancipatory transformations.

While the political art of staging equality in public space is a vital moment, the process of transformation requires indeed the slow but unstoppable production of new forms of spatialization quilted around materializing the claims of equality, freedom, and solidarity. In other words, what is required now and what needs to be thought through is if and how these proto-political localized events can turn into a spatialized political 'truth' procedure; a process that has to consider carefully the persistent obstacles and often violent strategies of resistance orchestrated by those who wish to hang on to the existing state of the situation. This procedure also raises the question of political subjectivation and organizational configurations, and requires perhaps forging a political name that captures the imaginary of a new egalitarian commons appropriate for 21st century's planetary form of urbanization. While during the 19th and 20th century, these names were closely associated with 'communism' or 'socialism' and centred on the key tropes of the party as adequate organizational form, the proletarian as privileged political subject, and the state as the arena of struggle and site to occupy, the present situation requires a re-imagined socio-spatial configuration and a new set of strategies that nonetheless still revolve around the notions of equality. However, state, party, and proletarian may not any longer be the key axes around which an emancipatory sequence becomes articulated. While the remarkable uprisings since 2011 signalled a desire for a different political configuration, there is a long way to go in terms of thinking through and acting upon the modalities that might unleash a transformative democratic political sequence. Considerable intellectual work needs to be done and experimentation is required in terms of thinking through and pre-figuring what organizational forms are appropriate and adequate to the task. What is the terrain of struggle, and what or who are the agents of its enactment?

The urgent tasks now to undertake for those who maintain fidelity to the political events choreographed in the new insurrectional spaces that demand a new constituent politics (that is a new mode of organizing everyday urban life) revolve centrally around inventing new modes and practices of collective and sustained political mobilization, organizing the concrete modalities of spatializing and universalizing the Idea provisionally materialized in these intense and contracted localized insurrectional events and the assembling of a wide range of new political subjects who are not afraid to stage an egalitarian being-in-common, imagine a different commons, demand the impossible, perform the new, and confront the violence that will inevitably intensify as those who insist on maintaining the present order realize that their days might be numbered. Such post-capitalist politics is not and cannot be based solely on class positions.

The aftermath of the insurgencies of the past few years saw a veritable explosion of new socio-spatial practices that are experimented with, from housing occupations and movements against dispossession in Spain to rapid proliferation of experimenting with new egalibertarian life-styles and forms of social and ecological organization in Greece, Spain and many other places, alongside more traditional forms of political organizing (for example, DIEM2025 in Europe or Podemos in Spain). Not all experimentations will succeed. Many will fail. In the face of inevitable setbacks, the fidelity to the democratizing process needs to be maintained and sharpened. An extraordinary experimentation with dispossessing the dispossessor, with reclaiming the commons and organizing access, transformation, and distribution in more egalibertarian ways already marks the return to 'ordinary' life in the aftermath of the insurgencies. The incipient ideas expressed in the event are materialized in a variety of places and ways, and in the midst of painstaking efforts to build alliances, bridge sites, repeat the insurgencies, establish connectivities and, in the process, produce organization, symbolize its practices, and generalize its desire. While the political art of staging equality in public squares is a vital moment in re-ordering 'the partition of the sensible,' the process of transformation requires the slow but unstoppable production of new forms of spatialization quilted around materializing the claims of equality, freedom, and solidarity. This is the promise of the return of the political embryonically performed in the artistic violence of insurgent urban practices.

Notes

1 The second part of the chapter is based on Wilson, Japhy and Swyngedouw, Erik (2014) 'Insurgent Architects, Radical Cities and the Promise of the Political,' in Wilson J. and E. Swyngedouw eds., *The Post-Political and its Discontents: Spaces of Depoliticization, Specters of Radical Politics.* Edinburgh: Edinburgh University Press, 169–188.
2 See www.politicalbeauty.de/mauerfall.html (Accessed 15.01.2015).

Bibliography

Badiou, A. 2003. 'Fifteen Theses on Contemporary Art.' In *The Drawing Centre*, 4.2003. Available at: www.lacan.com/issue22.php (Accessed 15.01.2015).

Badiou, A. 2005. *Being and Event*. London: Continuum.

Badiou, A. 2006. *Metapolitics*. London: Verso.

Badiou, A. 2012. *The Rebirth of History: Times of Riots and Uprisings*. London: Verso.

Dikeç, M. and Swyngedouw, E. 2017. 'Theorizing the Politicizing City.' In *International Journal of Urban and Regional Research* 41(1), 1–18.

Foucault, M. 2007. *Security, Territory, Population: Lectures at the Collège de France 1977–1978*. London: Palgrave Macmillan.

García Lamarca, M. 2013. 'Insurgent practices and housing in Spain: Making Urban Commons?' Paper presented at Symposium *Urban Commons: Moving beyond State and Market*, George Simmel Centre for Metropolitan Research, Humboldt University, Berlin, 27–28.09.2013.

Harvey, D. 2000. *Spaces of Hope*. Edinburgh: Edinburgh University Press.

Harvey, D. 2012. *Rebel Cities – From the Right to the City to the Urban Revolution*. London: Verso.

May, T. 2008. *The Political Thought of Jacques Rancière-Creating Equality*. Edinburgh: Edinburgh University Press.

Mouffe, C. 2005. *On the Political*. London: Routledge.

Rancière, J. 1998. *Disagreement*. Minneapolis: University of Minnesota Press.

Rancière, J. 2000. *Le Partage du Sensible: Esthétique et Politique*. Paris: La Fabrique.

Rancière, J. 2001. 'Ten Theses on Politics.' In *Theory & Event* 5(3). Available at: https://muse.jhu.edu/article/32639 (Accessed 10.07.2019).

Rancière, J. 2003. 'Politics and Aesthetics: An Interview.' In *Angelaki- Journal of the Theoretical Humanities* 8(2), 194–211.

Swyngedouw, E. 2013. 'Where is the Political? Insurgent Mobilizations and the Incipient "Return of the Political."' In *Space and Polity* 18(2), 122–136.

Swyngedouw, E. 2018. *Promises of the Political*. Cambridge, MA: The MIT Press.

Wilson, J. and Swyngedouw, E. eds. 2014. *The Post-Political and its Discontents: Spaces of Depoliticization, Specters of Radical Politics*. Edinburgh: Edinburgh University Press.

Žižek, S. 2013. *Demanding the Impossible*. Cambridge: Polity Press.

8 Athens, invisible city of the 21st century

From Olympic illusions to crisis and its contestation by the urban grassroots

Lila Leontidou

'Invisible' and 'soft' cities in cultural geography

'Invisible cities' of the mind, as Italo Calvino (1974) named them for posterity, were always there, behind material ones; but their legitimation within Geography and Urban Studies came after May 1968 and the revolutionary upheavals which transformed sciences as well. Geography then passed from the quantitative revolution to post-positivist humanistic, radical, and critical approaches: hermeneutics, mental maps, intersubjectivity, symbolism, including an interaction and conversation with literature. Imagination, memory, and a reflective attitude weighed perhaps more than hard-data science or exclusive reliance on empirical reality (Leontidou 2011). Soon mental maps, 'geographical imaginations' (Gregory 1994), intersubjectivity, and 'cities of collective memory' (Boyer 1996) would become a key aspect of understanding spatialities.

Inversely, the early 1970s also saw works of literature with space as their main concern: first, Calvino's poetic book *Invisible Cities* opened up a large territory of connections of urban space with memory, imagination, culture, art, desire. He wrote:

> Cities, like dreams, are made of desires and fears, even if the thread of their discourse is secret, their rules are absurd, their perspectives deceitful, and everything conceals something else.
>
> (Calvino 1974, 44)

And:

> This said, it is pointless trying to decide whether Zenovia is to be classified among happy cities or among the unhappy. It makes no sense to divide cities into these two species, but rather into another two: those that through the years and the changes continue to give their form to desires, and those in which desires either erase the city or are erased by it.
>
> (Calvino 1974, 35)

In 1974, the year when this was translated from the Italian, a British travel writer, Jonathan Raban, presented his own book, *Soft City*, for cities of imagination, dream, art, as distinct from 'hard cities':

> The city as we imagine it, the soft city of illusion, myth, aspiration, nightmare, is as real, maybe more real, than the hard city one can locate on maps in statistics, in monographs on urban sociology and demography and architecture.
>
> (Raban 1974, 4)

The synchronicity of the two books has inspired this essay, which reflects on the material world as affected by interpretation, memory, and interaction with the invisible world, which it then reshapes, in an eternal spiral. This spiral has long affected Athens, a city at the crossroads of many epochs and different worlds from antiquity and throughout space-time – between occident and orient, at the intersection of three continents and several civilizations (Leontidou and Martinotti 2014). We will attempt here a conversation with Italo Calvino, in the way he taught us to converse and interact, with the encounter between Marco Polo and Kublai Khan. In the core of the 55 imaginary cities in this inspired book, there is Marco Polo's home city, Venice:

> The emperor did not turn a hair. 'And yet I have never heard you mention that name.'
>
> And Polo said: 'Every time I describe a city I am saying something about Venice.'
>
> Calvino 1974, 86–87)

His book is a conversation between people, while our essay is one between cities, Venice and Athens, wherein the mystery is reversed: Calvino keeps masking Venice, while our reader knows the starting point, Athens; the surprise remains, however, in the succession of unexpected 'invisible cities' engulfed in the different facets of Athens. This city has become a constellation and simultaneity of century-old stories, combining ancient and modern, original and copy, informal as postmodern (Leontidou 1993), theatricality and squalor, tourism and romantic travelling, migration and motionless poverty, throughout its history. This is still going on during the most pronounced liquid modernity (Bauman 2000) in the short 21st century, with a rapid succession of invisible cities in the course of only two decades.

Greater Athens entered the new millennium with 3,187,734 inhabitants (2001), reduced to 3,122,540 during the crisis (2011) according to ELSTAT temporary data. This short 21st century was also characterized by a rapid succession or occasional coexistence of eutopia and dystopia. Along the centuries, Athens has always emerged from calamity as a city of hope (Leontidou 2020), giving its form to desires, imaginations and intersubjective narratives. But there were challenging periods when it was less so, and one of them is quite recent, too. Its kaleidoscopic

transitions in the 21st century can be summed up as four phases of urban transformation: from **Olympic** illusions and their short-lived eutopia, we reach to the
city of **Crisis** and austerity, smudged by EU-inspired quasi-'Orientalism' (Said
1978) or 'crypto-colonialism' (Herzfeld 2002); but the city has proved **Resilient**
after networked urban social movements reached out to a version of the **Smart**
city of grassroots digital initiatives contesting the crisis. Every one of these facets
of Athens can be named, following Calvino's imagination.

Maurilia: Olympic illusions and the entrepreneurial city

[S]ometimes different cities follow one another on the same site and under the
same name, born and dying without knowing one another, without communication among themselves . . . It is pointless to ask whether the new ones are better or
worse than the old, since there is no connection between them, just as the old post
cards do not depict Maurilia as it was, but a different city which, by chance, was
called Maurilia, like this one.

(Calvino 1974, 30–31)

The Olympic Games, old and new, in Olympia or in Athens, are heavy with cultural symbolism and constitute a pillar of modern Greek identity: 'The Greek community unanimously regards the Olympics as a kind of spiritual child' (Kasimati
and Vagionis 2017, 156). Athens expressed pride for having revived the ancient
Olympics in 1896 in the celebrated 'Kallimarmaro' marble stadium. A hundred
years later, in 1996, a different city on the same site competed in the global arena
for the 'Golden Olympics.' Despite this unsuccessful bid, Athens kept insisting. It
had already started building new infrastructure including a new metro (since January 29, 2000) and international airport (with the first take-off on March 29, 2001).
It had irreversibly entered global neoliberal urban competition for mega-events,
clumsily adopting entrepreneurial city strategies of image promotion, place mar-
keting and megaprojects (Leontidou 2006).

Efforts to attract mega-events culminated in the phantasmagoria of the 2004
Olympics. Athens restructured its image with the familiar double strategy of valorization of ancient tradition and myth, especially with pedestrianization around
the archaeological sites (Loukaki 2008; Leontidou 2006); and new infrastructure
with innovative architecture and global urban design. The centralist state and its
'exceptional' legal framework (for environmental protection, building permits,
expropriations) cooperated with quangos, and although public–private partner-
ships (PPPs) were soon abandoned (Souliotis, Sayas and Maloutas 2014), such
neoliberal mixtures set in motion an unending spiral of urban redevelopment with
spectacular sporting venues, postmodern architecture and flagship projects, especially Santiago Calatrava's Olympic park (Figure 8.1).

International media and foreigners initially criticized Athens for inadequacy
in completing the works in time. However, they finally had to admit that the
mega-event was successful and started praising the Olympics which 'returned
to their birthplace' just over a century after their revival in the same city. Athens

Figure 8.1 The Olympic Stadium in Athens by Santiago Calatrava with the Olympic flame alight. Photo shot during the third day of the Athens Olympics, 2004.

Source and copyright: Lila Leontidou.

thus had its moment of glory in 2004, but to the sacrifice of the compact city, sustainability, and economic viability. Urban expansion was reinforced and urban sprawl engulfed eastern Attica, as infrastructure expanded and Olympic works affected previously uninhabited areas (Couch et al. 2007; Chorianopoulos and Pagonis 2019).

The eutopia was also shadowed by contradictions in civil society. Athens did not achieve a broader socio-political civic consensus in the model of Barcelona in 1992. On the one extreme, a minority opposed the fiestas and overspending, some citizen movements and collectivities were disappointed and planners objected to the location of venues. However, popular contestations were weak and hesitant, and, on the other extreme, public opinion polls indicated a majority of pride (Kasimati and Vagionis 2017, 154; Panagiotopoulou 2014) and intersubjective sense of success, while volunteer citizen participation was significant, as

> the Athenian urban majority accepted that the so-called Olympic culture can be sold to an obscure global community, whereas in fact it was this community . . . that has sold it to the Athenian public.
>
> (Afouxenidis 2006, 292)

The Greek government and elites celebrated the return of the Olympic Games to their 'cradle,' Greece, but in fact, the Olympic 'soft city' was mainly constructed abroad. Athens thus had its moment in global urban competition and found its positive niche in globalization as an unrecognizable 'Maurilia,' imagining that the Olympic 'success story' would be long lasting. But the eutopia was short lived, following the fate and the peculiar function of pomp and monumentalization in Greek history. Overspending in the midst of decline has been a centuries-old counteractive strategy in Greece. Olympic illusions in periods of crisis, bankruptcy, and defeat include the revival of the ancient Olympic Games in the 19th century, which was staged between the national bankruptcy of 1893 and the humiliating Greco-Turkish war, which followed the resurgence of the 'Great Idea' of Greek irredentism and illusions of grandeur and led to a humiliating defeat by the Ottoman Empire in 1897 (Leontidou 1989). The 1896 Olympics were a temporary confirmation of dignity and an intersubjective reassurance for citizens seeking solace and identity. Declining 'hard cities' would strive to create imaginaries of thriving 'soft cities,' by staging major events and constructing luxurious monuments in the urban landscape. The same happened in Hermoupolis, Syros – but that is yet another long story.[1]

Could this be the case for the 2004 Olympics as well? After their end, instead of valorization, a spectre of bankruptcy haunted Greece. The costs and benefits of the Olympics have never been properly assessed, and the final cost has been a well-guarded secret (Panagiotopoulou 2014, 176–180). It seems that the positive impact on employment creation and cultural tourism lasted during the 1997–2005 period (Kasimati and Vagionis 2017, 150). Then the drain of public funds and post-Olympic under-use of Olympic venues, many of which fell into disrepair, counterbalanced the benefits. In fact, the Olympics 'failed to create the institutional conditions for their extended reproduction' (Souliotis et al. 2014) and became a catalyst for the implementation of neoliberal urban policies. 'Maurilia' was transformed into a city with infrastructural 'white elephants' burdening the Greek taxpayer and contributing to the economic crisis (Kasimati and Vagionis 2017, 160). The dystopia culminated during the debt crisis of the 2010s, when disaster struck, misery spread throughout the country, and unrelenting privatizations devoured urban assets.

Theodora: neoliberal crisis, crypto-colonialism, and the city

> Recurrent invasions racked the city of Theodora in the centuries of its history; no sooner was one enemy routed than another gained strength and threatened the survival of the inhabitants. When the sky was cleared of condors, they had to face the propagation of serpents.
>
> (Calvino 1974, 159)

Successive enemies within and around Southern Europe gained strength and were then defeated. The region had barely managed to rid itself of dictatorships in the

1970s and enjoy the EU Southern enlargement in the 1980s, when a new calamity engulfed it: the debt crisis. Shifting 'power-geometries' (Massey 2005) within the EU and vis-à-vis the rest of the Mediterranean reinforced virtual geopolitical borders, up to the reconstruction of a North/South divide within EU limits, within the Eurozone (Leontidou 2012a). This divide was sealed with bailout agreements, the so-called memoranda (*mnemonia*) for the Greek sovereign debt, and was combined with narratives of scorn and vilification of the South: a *déjà vu* of past Northern stereotypes as criticized by Antonio Gramsci in his account of the Mezzogiorno vilified in the 1920s as a 'ball and chain' obstructing the development of Northern Italy. The South was castigated for 'the organic incapacity of the inhabitants, their barbarity, their biological inferiority' (Gramsci 1971, 71; see Leontidou 2012a, 2014) by the North, which in fact exploited it. A century later, during the 2010s, we witnessed a similar North/South divide in the broader EU combined with the stigmatization of the South, originating in the North, especially Germany. Dystopic 'invisible cities' were constructed by colonial intersubjectivities and geographical imaginations (Gregory 1994). Not only did the European periphery as a whole suffer from neoliberal hard policies of austerity and 'accumulation by dispossession' (Harvey 2003), but it was also stigmatized as responsible for the crisis in a surge of disciplinary neoliberalism, intersubjectively constructing its own 'invisible cities.'

Northern elites, governments and money lenders have imposed on Athens their neoliberal policies through the Troika (IMF, ECB, EU). This was facilitated through branding the population as 'lazy' and corrupt, in narratives and abhorrent acronyms like so-called PIIGS, that is, the countries of the EU periphery. Contemporary Oriental views were not about 'languages, literatures, history, sociology, and so forth' (Said 1978, 206), but about debt and the crisis. Economics and finance have opened up a spectacle of hedge funds, spreads, interest rates, and inferior cultures (Leontidou 2014), which composed the 'invisible cities' of our times. The media went along. Such discourse, bordering on racism, burdened the South with responsibilities for the crisis and the collapse of the euro.

Stereotypes about a dystopic 'invisible city' justified the plundering of Athens and attempted to rob citizens of their dignity, to weaken them in order to accept austerity and its memoranda. The spectacle has even become visible in German periodicals' covers, and it has been extended to the UK and tourism: even the *Guardian* has gone as far as proposing 'poverty tourism' packages to crisis-hit places and refugee camps of Greece, turning misery to spectacle and commodity.[2] This was the shock doctrine of our times (Klein 2007) and the culmination of the most hideous Oriental and crypto-colonial phase in Europe, which must be sharply criticized as responsible for the crisis itself (Leontidou 2015).

In this dismal auste-city (Cappuccini 2017), the 'Theodora' of the 2010s, surrounded by enemies, the 'soft city' of EU power elites has affected the 'hard city' in adverse ways. Poverty escalated (Arapoglou 2019), urban degeneration and blight worsened by population loss combining disurbanization, a brain drain, destitute refugee crowds coming from war-stricken regions and living in substandard conditions (Maloutas 2019), deindustrialization and even detertiarization. Protests erupted by a mobilized youth practicing 'urbicide,' activists protesting, crowds in the piazzas. These were branded 'delinquents' or even 'terrorists'; but their own Athens has been a diametrically different 'invisible city,' as follows.

Anastasia: resilient city of spontaneous social movements and international solidarities

> For while the description of Anastasia awakens desires one at a time only to force you to stifle them, when you are in the heart of Anastasia one morning your desires waken all at once and surround you. The city appears to you as a whole where no desire is lost and of which you are a part.
>
> (Calvino 1974, 12)

'Anastase' is the Greek word for 'resurrection' and echoes the description of Greece as the center of an 'anarchist renaissance' in Europe (Arampatzi and Nicholls 2012). One day in late May 2011, the desires and imaginations of citizens awakened all at once, and Athens came to life. Urban piazzas were often theatres of spontaneous movements and uprisings, culminating in the 1970s in the Exarcheia neighbourhood around and in the Polytechnic University, where the student uprising brought down a junta. The clashes on November 17, 1973, are celebrated until today. But now we have reached the digital epoch, the hybrid city (Leontidou 2015), and social movements have changed in important ways (Leontidou 2010; Vagionis 2019).

Syntagma Square has seen several mobilizations through centuries, culminating in the fragmented and violent spontaneous riots of December 2008, after a young pupil was shot by the police (Vradis and Dalakoglou 2011). Then Syntagma awakened for good in the summer of 2011, when the Spaniards from the Puerta del Sol in Madrid probed the Greeks: 'wake up'! This push was unnecessary, because popular indignation for austerity and suppression had peaked. The Athens 'movement of the piazzas,' with nuances and divisions in its midst (Leontidou 2012b, 2014; Arampatzi and Nicholls 2012; Kaika and Karaliotas 2016; Kavoulakos 2019), was extended with protests for the closure of the Hellenic TV in 2013 (Leontidou 2014).

Grassroots 'movements of the piazzas' shook the broader Mediterranean in 2011–13 and lit up the night, literally, with the screens of cell phones and Skype windows, and figuratively, with their *joie de vivre* (Leontidou 2014, 2020). Citizens contested the hierarchical power geometries of disciplinary neoliberalism and the heavy-handed strategies of 'accumulation by dispossession.' They strove to recapture their dignity during and after the movements, because they knew they had nothing but austerity to expect from their governments and the EU on the top of the hierarchy. International networking in spontaneous popular assemblies and various cultural events is exemplified by conversations between Syntagma and Puerta del Sol, Madrid, by Skype, which have empowered both cities in the summer of 2011 (Leontidou 2012b, 308).

Figure 8.2 The puppet representing Sophocles' Antigone as well as Justice, in the performance by the Theatre du Soleil in Syntagma during the movement of the piazzas in 2011, directed by Ariane Mnouchkine, who was also present.

Photo source and copyright: L. Leontidou. Photo shot in Syntagma Square.

In this unprecedented domino effect of anti-austerity social movements in the
Mediterranean after 2011, resilient cities awakened. Policy makers stretch this
recent buzzword beyond natural disasters, to encompass the urban economic crisis and social misery (Fainstein 2014). Having resilience requires accommodating
to unavoidable events which produce system change, not preventing them. Since
they result from the interaction of multiple, uncoordinated factors, no public agent
has the power to control them. Only spontaneous action can confront them. And,
as it is, the Mediterranean grassroots has always lived in spontaneity as Gramsci
(1971, 196) defined it. From the refugee inflow of the 1920s until the postwar
period, semi-squatter settlements, 'slums of hope,' informal work and small enter-
prises created a resilient urban landscape of mixed land use (Leontidou 1990).

Spontaneous uprisings followed in the 2010s, countering austerity, poverty, and
all top-down strategies of quasi-Orientalist stigmatization of any initiative (Leontidou 2012b). Resilience now meant alternative grassroots life patterns in the piazzas and work practices after the mobilizations, as well as support by international
solidarity networks, which kept pouring in. Several creative and talented people,
collectivities and celebrities expressed on site the attitudes, perceptions, intersubjectivities and geographical imaginations of a large global community from the rest of

Figure 8.3 Multi-activity mobilization with concert stage and public discussion in 2013, protesting the abrupt closure of the Hellenic TV. From left to right, Dimitris Dalakoglou translating to David Harvey, Kostas Douzinas, Aglaia Kyritsi.

© Photo and source: L. Leontidou. Photo shot in the Hellenic TV courtyard.

Europe and the Americas. They paid homage to the contribution of Mediterranean cities to global culture and looked ahead to new hopeful resilient cities of postcapitalism (Mason 2016). Athens hosted several collectivities and intellectuals offering sympathy and stirred the imagination of artists acknowledging their indebtedness to classical Greece and its heritage – like the Theatre du Soleil, performing Sophocles' Antigoni in Syntagma (Leontidou 2012b, 305) (Figure 8.2).

Activists and intellectuals came together to counter autocracies and demand direct democracy. The undercurrent of spontaneity, combined with the use of the Internet and social media, awakened people and instantly brought them to the streets, not of one, but of several cities (Castells 2012; Robins 2019). The 'invisible city' corresponded to the imaginations of artists, academics, activists, who arrived to talk and act in solidarity with the mobilized grassroots, to stage their performances and concerts in the piazzas and to expose quasi-Orientalism and crypto-colonialism. They kept assembling till 2013 in the piazzas, on the streets, in the Hellenic TV courtyard. Mobilizations in the heart of 'Anastasia,' awakened desires for freedom in Athens during 2011–13 (Figure 8.3).

Democratic traditions went through several backlashes and adventures in the 2010s, with technocrats appointed in electable posts, even that of the prime minister; but the massive and spontaneous movements persisted (Leontidou 2014; Vradis 2019). And in January 2015 they finally made a difference in politics: they led to the electoral victory of the Left, Syriza, even if this was not destined to rise to the occasion later. At the time, the Left in government heightened international solidarity and mobilized the sizeable Greek diaspora of the brain drain, which keeps going until today. Athens has been resilient, in the sense of accommodating to unavoidable events like neoliberal austerity, not preventing them. It belongs to a new generation of 'invisible cities' of the South, resilient in countering the Northern power elites and challenging their TINA (There Is No Alternative) with TAPA (There Are Plenty of Alternatives), seeking solidarity, networking, resistance, and direct democracy.

Tamara: the grassroots 'smart city'

> Your gaze scans the streets as if they were written pages: the city says everything you must think, makes you repeat her discourse, and while you believe you are visiting Tamara you are only recording the names with which she defines herself and all her parts.
>
> (Calvino 1974, 14)

The digital society in the 21st century has created another sort of city, virtual and material at the same time, or a hybrid city (Leontidou 2020). Calvino's cities were invisible in the sense of being cities of the mind, but today's cities are invisible also because they are registered in virtual space, with sorts of text and hypertext creating hybridity: broadening their spaces toward remote cities, representing global urban experience as well as the here and now. The online 'invisible city' unites with the mental one, creating a hybrid 'invisible city' in liquid modernity (Bauman 2000).

The urban social movements of the 2010s had important consequences for politics, of course, but also for urban transformation, coupled with technology, ICT, and the power of networking (Castells 2012). Alternative activities and art forms emerged in the context of solidarity networks rather than power hierarchies, virtually as well as materially, by collectivities and the grassroots rather than the power elite, informally, as precariously as the graffiti and as alternatively as site-specific art and performance. And when crowds withdrew from the piazzas, their invisible online networks kept working. New hubs of creativity sprang up in several corners of the city. The Social and Solidarity Economy (SSE) and the 'Commons' became the new buzzwords during the crisis of the 2010s (Stavrides 2016; Leontidou 2017; Kavoulakos 2019; Portaliou 2019).

These might well be the beginnings of postcapitalism (Gibson-Graham 2006; Mason 2016):

> New forms of ownership, new forms of lending, new legal contracts: a whole business subculture has emerged over the past ten years, which the media has dubbed the 'sharing economy.' Buzz-terms such as the 'commons' and 'peer-production' are thrown around, but few have bothered to ask what this means for capitalism itself. . . . [T]his is no longer my survival mechanism, my bolt-hole from the neoliberal world, this is a new way of living in the process of formation.
>
> (Mason 2016, xv)

In urban hybrid public and common spaces, the virtual and the material mingle. The digitization of everyday life constitutes the most democratic industrial revolution ever, with the means of production freely available in everyone's hands: smartphones, laptops, tablets. On the face of digital popular initiatives, in a previous paper (Leontidou 2015) we have challenged mainstream, neoliberal conceptualizations of the 'Smart City,' by inserting grassroots creativity and solidarity into their logic. In theory, this concept can be recuperated from the neoliberal narrative and disassociated from power elites. In the midst of austerity, popular creativity combines technology and knowledge of an educated young population familiar with ICT use to rejuvenate the notion of the 'Smart City.' In Athens, but also in other Greek towns and cities, like Trikala, the debt crisis has paved the way for a grassroots version of the 'Smart City.'

A typology for several types of diverse collectivities, alternative economies, and productive initiatives (Leontidou 2015, 87–94) reveals an oxymoron: that the debt crisis is paving the way for grassroots creativity following popular mobilizations. Networks expand beyond national borders. Based on ICT and the unemployed or underemployed educated young labour force, Mediterranean cities are probably becoming cosmopolitan again, in the sense of our epoch, when cosmopolitanism is virtual as well as material, and can move rapidly from online to offline forms of creativity and vice versa, simultaneously among several cities.

Top-down policy keeps focusing on misery, with handouts to the poor, but gradually grassroots inventiveness, digital popular creativity and resilience are appreciated by the Syriza government until 2019, with support, dissemination, and education (Leontidou 2020). 'Tamara' with her hypertexts will supersede neoliberal stereotypes, as a grassroots version of the Smart City in the near future. Such a version tends to become reality in Greece, and the invisible city of digital democracies is bound to affect the visible one soon, as a probable way out of the dystopia.

Clarice: a final comment on eutopia and dystopia

> Clarice, the glorious city, has a tormented history. Several times it decayed, then burgeoned again, always keeping the first Clarice as an unparalleled model of

every splendor, compared to which the city's present state can only cause more sighs at every fading of the stars.

(Calvino 1974, 106)

Athens IS 'Clarice,' with antiquity as its remote unparalleled model of splendor, complicating the present with dystopias and eutopias, which are not just 'objective,' nor do they appear in a linear sequence. They belong to 'invisible cities' imagined by different social actors in and out of the country, in different ways, as we demonstrated in this essay. The power elites' dystopias clash with utopias of activists and solidarity agents. Athens has been stigmatized in the quasi-Oriental discourse of crypto-colonial Northern European elites and their media, but praised and followed by intellectuals, tourists, collectivities and solidarity networks, materially or online. For all its history, urban eutopias and dystopias, hope and nightmare, alternate, coexist and interpenetrate each other within one and the same city, Athens; or Venice, just as in Calvino's book inspiring this essay, as Raban (1974, 13) explicitly states:

utopias and dystopias go, of necessity, hand in hand. Disillusion is a vital part of the process of dreaming – and may, one suspects, prove almost as enjoyable.

In simultaneity with Raban, Calvino also builds upon this duplicity. The coexistence of eutopias and dystopias culminates in Beersheba, the 'invisible city' where heavenly perfection is believed to be built upon a filthy underground city, but this is not unrelenting dystopia either:

The inferno that broods in the deepest subsoil of Beersheba is a city designed by the most authoritative architects, built with the most expensive materials on the market, with every device and mechanism and gear system functioning, decked with tassels and fringes and frills hanging from all the pipes and levers.

(Calvino 1974, 112)

This duplicity, inspired from Venice, definitely brings the Olympic Athens to mind, an eutopia that fell to disrepair; but this also portrays the city of Athens diachronically: built by brilliant architects as an ancient and a neoclassical and a late modern city, with treasures often buried below heaps of filth, this iconic city has enchanted personalities from antiquity, then Roman emperors like Hadrian, who built monuments, an aqueduct, and protected the city (Leontidou and Martinotti 2014), and then again stirred the imagination of European intellectuals: Gustav Flaubert, Sigmund Freud, Simone de Beauvoir, modern artists like Le Corbusier, and postmodern intellectuals like Jacques Derrida, who kept exploring it. European identity has largely drawn from Hellenism through selective reinterpretations of classical Athens,[3] but even now, when construction gave way to deconstruction or even destruction, in imagination but also materiality, contradictions abound. European identity is negotiated in this 'invisible city' between elites exerting crypto-colonial domination, and collectivities in solidarity, promising that revival is around the corner in the city of hope (Leontidou 2020). This is taking place in the course of a discontinuous and very uneven urban history, a tiny fragment of which has been discussed here.

Notes

1 The city of Hermoupolis in the island of Syros spent heavily on civic design and a decorous municipal building in the 1890s, as soon as it sensed that decay was close and depopulation toward Athens and Piraeus had set in (Leontidou 1989). This effort did not prevent the decline, but it was a reassurance for the population.
2 See www.rt.com/news/422572-guardian-refugee-tourism-greece/ (Accessed 16.01.2019) and the actual ad on March 28, 2018, together with Facebook criticism. The ad was withdrawn a few days later, due to trolling on Facebook, Twitter, and the Internet in general. As to the German magazines' offensive covers, they will not be shown here.
3 Such as the neoclassical utopia in the 18th century, which the Bavarians in the 19th century returned into Athens, the city which had inspired it (Boyer 1996; Loukaki 2008; Leontidou and Martinotti 2014) as if to confirm Hellenism in its cradle. However, the same city later demolished much of its neoclassical heritage (Leontidou 1990), and most recently turned to a dystopia during the crisis.

Bibliography

Afouxenidis, A. 2006. 'Urban Social Movements in Southern European Cities: Reflections on Toni Negri's "The Mass and the Metropolis".' In *City* 10(3), 287–294.

Afouxenidis, A. ed. 2012. *Anisoteta sten Epoche tes Kriseos: Theoretikes kai Empeirikes Dierevneseis (Inequality in the Period of Crisis: Theoretical and Empirical Investigations)*. Athens: Propobos.

Afouxenidis, A., Gialis, S., Iosifides, T. and Kourliouros, E. eds. 2019 (in English and Greek). *Geographies sten Epoche tes Refstotetas (Geographies in a Liquid Epoch)*. Athens: Propobos.

Arampatzi, A. and Nicholls, W.J. 2012. 'The Urban Roots of Anti-neoliberal Social Movements: The Case of Athens, Greece.' In *Environment and Planning A* 44, 2591–2610.

Arapoglou, V. 2019. 'Greek Cities at the Thresholds of Neoliberalism. Translations of Urban Anti-poverty Policies in Southern Europe.' In Afouxenidis, T. et al. eds. 2019, 206–222.

Bauman, Z. 2000. *Liquid Modernity*. Cambridge: Polity Press.

Boyer, M.C. 1996. *The City of Collective Memory: The Historical Imagery and Architectural Entertainments*. Cambridge, MA: The MIT Press.

Calvino, I. 1974. *Invisible Cities*, translated by Weaver, W. London: Harvest and HBJ.

Cappuccini, M. 2017. *Austerity and Democracy in Athens: Crisis and Community in Exarcheia*. Cham: Palgrave Macmillan.

Castells, M. 2012. *Networks of Outrage and Hope: Social Movements in the Internet Age*. Cambridge: Polity Press.

Chorianopoulos, I. and Pagonis, T. 2019. 'Synecheies kai Asynecheies sten Poreia tes Mesogeiakes Poles: Opseis tes Astikes Diachyses kai Politikes sten Athena (Continuities and

Discontinuities in the Itinerary of the Mediterranean City: Facets of Urban Sprawl and Policy in Athens).' In Afouxenidis, T. et al. eds. 2019, 276–294.

Couch, C., Leontidou, L. and Petschel-Held, G. eds. 2007. *Urban Sprawl in Europe: Landscapes, Land-use Change, and Policy*. Oxford: Blackwell.

Fainstein, S. 2014. 'Resilience and Justice.' In *International Journal of Urban and Regional Research* 38, 157–167.

Gibson-Graham, J.K. 2006. *A Postcapitalist Politics*. Minneapolis: University of Minnesota Press.

Gospodini, A. and Beriatos, H. eds. 2006. *Ta Nea Astika Topia kai he Hellenike Pole (The New Urban Landscapes and the Greek City)*. Athens: Kritike.

Gramsci, A. 1971. *Selections from the Prison Notebooks*, edited by Hoare, G. and Smith, G.N. New York: International Publishers.

Gregory, D. 1994. *Geographical Imaginations*. Oxford: Blackwell.

Harvey, D. 2003. *The New Imperialism*. Oxford: Clarendon Press.

Herzfeld, M. 2002. 'The Absent Presence: Discourses of Crypto-Colonialism.' In *The South Atlantic Quarterly* 101(4), 899–926.

Kaika, M. and Karaliotas, L. 2016. 'The Spatialization of Democratic Politics: Insights from Indignant Squares.' In *European Urban and Regional Studies* 23(4), 556–570.

Kasimati, E. and Vagionis, N. 2017. 'Cultural Tourism and the Olympic Movement in Greece.' In Carson, S. and Pennings, M. eds., *Performing Cultural Tourism: Communities, Tourists and Creative Practices*. London: Routledge, 147–163.

Kavoulakos, K.I. 2019. 'To Dikaioma sten Pole: Koinonika Kinemata sten Athena tou 21ou Aiona (The Right to the City: Social Movements in Athens of the 21st Century).' In Afouxenidis, T. et al. eds. 2019, 327–342.

Klein, N. 2007. *The Shock Doctrine: The Rise of Disaster Capitalism*. New York: Metropolitan Books, Henry Holt & Co.

Knox, P. ed. 2014. *Atlas of Cities*. Princeton: Princeton University Press.

Leontidou, L. 1989 (in Greek with summary in English). *Cities of Silence: Working-Class Space in Athens and Piraeus, 1909–1940*. Athens: ETVA/PIOP.

Leontidou, L. 1990. *The Mediterranean City in Transition: Social Change and Urban Development*. Cambridge: Cambridge University Press.

Leontidou, L. 1993. 'Postmodernism and the City: Mediterranean Versions.' In *Urban Studies* 30(6), 949–965.

Leontidou, L. 2006. 'Diapolitismikoteta kai Heterotopia sto Mesogeiako Astiko Topio: Apo ten Afthormite Astikopoiise sten Epiheirematiki Pole (Inter-culturalism and Heterotopia in the Mediterranean Cityscape: From Spontaneous Urbanisation to the Entrepreneurial City).' In Gospodini, A. and Beriatos, H. eds. 2006, 70–84.

Leontidou, L. 2010. 'Urban Social Movements in "Weak" Civil Societies: The Right to the City and Cosmopolitan Activism in Southern Europe.' In *Urban Studies* 47(6), 1179–1203.

Leontidou, L. 2011. *Ageografetos Chora: Ellenika Eidola stous Epistemologikous Anastohasmous tes Evropaikes Geografias (Geographically Illiterate Land: Hellenic Idols in the Epistemological Reflections of European Geography)*, Rev. 8th ed. Athens: Propobos.

Leontidou, L. 2012a. 'He Anakataskeve tou "Evropaikou Notou" sten Meta-apoikiake Evrope: Apo ten Taxike Syggrouse stis Politismikes Taftotetes (The Reconstruction of the "European South" in Post-Colonial Europe: From Class Conflict to Cultural Identities).' In Afouxenidis, A. ed. 2012, 25–42.

Leontidou, L. 2012b. 'Athens in the Mediterranean "Movement of the Piazzas": Spontaneity in Material and Virtual Public Spaces.' In *City* 16(3), 299–312.

Leontidou, L. 2014. 'The Crisis and its Discourses: Quasi-Orientalist Attacks on Southern Urban Spontaneity, Informality and *Joie de Vivre*.' In *City* 18(4–5), 546–557.

Leontidou, L. 2015. ' "Smart Cities" of the Debt Crisis: Grassroots Creativity in Mediterranean Europe.' In *The Greek Review of Social Research* 144(A), 69–101.

Leontidou, L. 2017. 'Commoning in the 21st-Century City.' In *City* 21(6), 902–906.

Leontidou, L. 2020. 'Mediterranean Cities of Hope: Grassroots Creativity and Hybrid Urbanism Resisting the Crisis.' In *City* 24(1). doi:10.1080/13604813.2020.1739906

Leontidou, L. and Martinotti, G. 2014. 'The Foundational City.' In Knox, P. ed., *Atlas of Cities*. Princeton: Princeton University Press, 16–33.

Loukaki, A. 2008/2016. *Living Ruins, Value Conflicts*. London: Routledge (first edition 2008: Aldershot: Ashgate).

Maloutas, T. 2019. 'He These ton Metanastevtikon Omadon sten Koinonia kai sto Choro tes Athenas (The Place of Migrant Groups in the Society and Space of Athens).' In Afouxenidis, A. et al. eds. 2019, 225–250.

Mason, P. 2016. *Postcapitalism: A Guide to our Future*. London: Penguin, Random House.

Massey, D. 2005. *For Space*. London: Sage.

Panagiotopoulou, R. (2014). 'The Legacies of the Athens 2004 Olympic Games: A Bitter-Sweet Burden.' In *Olympic Contemporary Social Science, Special Issue: The Olympic Legacy*, 9(2), 173–195.

Portaliou, H. 2019. 'Apo ten Allelegye Koinotikou kai Syntechniakou Charactera sta Syghrona Kinemata Koinonikes Antistases kai Allilegyes (From Solidarity of a Communal and Corporate Character to Contemporary Movements of Social Resistance and Solidarity).' In Afouxenidis, A. et al. eds. 2019, 306–325.

Raban, J. 1974/1998. *Soft City*. London: Panther, Harvill Press.

Robins, K. 2019. 'A Space for Virtuosity: On Cultural Politics in the City.' In Afouxenidis, T. et al. eds, 78–97.

Said, E. 1978. *Orientalism*. London: Pantheon Books.

Souliotis, N., Sayas, J. and Maloutas, T. 2014. 'Megaprojects, Neoliberalization, and State Capacities: Assessing the Medium-term Impact of the 2004 Olympic Games on Athenian Urban Policies.' In *Environment and Planning C: Government and Policy* 32(4), 731–745.

Stavrides, S. 2016. *Common Space: The City as Commons*. London: Zed Books.

Vagionis, N. 2019. 'A Journey in Time and Space of Power Structures and Urban Social Movements.' In Afouxenidis, A. et al. eds. 2019, 296–303.

Vradis, A. 2019. 'Urban Bridges over Troubled Waters.' In Afouxenidis, A. et al. eds. 2019, 160–167.

Vradis, A. and Dalakoglou, D. eds. 2011. *Revolt and Crisis in Greece: Between a Present Yet to Pass and a Future Still to Come*. London: AK Press and Occupied London.

9 Bodies in the city

Athenian street art and the biopolitics of the 'Greek Crisis'

Dimitris Plantzos

(The) Society must be defended

The Archaeological Society at Athens was founded in 1837 in what then was the newly-established capital of a newly-established nation-state. As claimed in the Society's website, Athens at the time was but 'a mere village calling itself a city' (Archaeological Society 2019).[1] According to its constituent act, the mission of the Archaeological Society was (and still is) to bring the love of things ancient to the heart of the new-born kingdom, and at the same time to assist the official Greek State in 'i) the discovery, collection, conservation, keeping, restoration, repair and scholarly investigation of the monuments of Antiquity located in Greece and the Greek lands [. . .];[2] ii) the study of the life of the ancients and the exploration of Byzantine and Medieval archaeology and art, and iii) the stimulation 'in our parts' of an interest in fine arts at large and the diffusion of the knowledge of the history of ancient and later art.'[3] The fledgling Greek state was thus allowed to find in archaeology its ideological foundations and the material evidence for its existence, as the development of a national – Hellenic, as opposed to 'Greek' – archaeology offered to the newly-established kingdom a welcome repertoire of symbols and a convenient toolbox of cultural strategies that were going to prove very useful in the years to come (Plantzos 2008a).[4]

How does one turn, though, a 'village' into a 'city'?[5] The Society's august body of savants seem to offer their own answer to the question through their architectural interventions to the city's human-made space. Their present-day headquarters, on the corner of Panepistimiou and Omirou Streets in the heart of Athens, were designed and built in 1958 by architect Ioannis Antoniadis as a replacement of an earlier, neoclassical building erected in the beginning of the 20th century based on designs by Ioannis Axelos. Antoniadis had originally chosen for his building a 'classicizing yet subtle' structure (National Hellenic Research Foundation 2019); while work was in progress however, the Society's governing body demanded the addition of some characteristic antique-like embellishments such as a monumental Ionic front porch as well as classicizing pilaster caps, cornices, stone balcony-brackets, and so on. More recently, in late 2015, the solemn building on Panepistimiou Street received a further addition: a forbidding metal railing guarding its haughty entrance, thus highlighting the divide between the space where one cultivates the arts through the study of the 'glory that was Greece,' from the contemporary Athenian dystopia (Figure 9.1). In other words, maintaining the distinction between the venerable, studious bodies of the nation's archaeologists, and the unwanted bodies of the beggars, the drug addicts, and the approximately 9,000 homeless Athenians peopling the city's pavements and back alleys (cf. Arapoglou and Gounis 2015).[6]

Figure 9.1 Athens, city centre; headquarters of the Archaeological Society (front entrance).

© Photo and source: Dimitris Plantzos (2016).

As a visual culture, as well as a technopolitical strategy, Neoclassicism has played a key role in organizing urban environments and the bodies inhabiting them in modernity, at least since the days of Georges-Eugène Haussmann, if not earlier (de Moncan 2012; Clark 1984, 23–78; cf. Bastéa 110–117, 144, 188). Over the years, neoclassical edifices have come to be associated with the regime, and at times – as in the case of the recent upheavals in Greece resulting from a decade of recession, social destabilization, and strict austerity – this has made them unpopular with the general public, to the point of their being vandalized by angry demonstrators as symbols of a cold-blooded establishment (e.g., Plantzos 2012; Loukaki, this volume). Authorities tend to react by enforcing further exclusion tactics, in fact reversing the ostensibly public and open character of the buildings themselves and the fundamental principles of the neoclassical narrative at large.

The Society's decision to deface its own building in order to cleanse its doorway from the miasma of Athens's *vitae nudae* is a typical example of the ways in which Greek life has turned into a new, and excruciatingly more violent, breed of biopolitics in recent years. For, if biopolitics is the style of government, or a 'set of mechanisms,' if you will, 'through which the basic biological features of the human species became the object of political strategy' – what Michel Foucault has called 'bio-power' (Foucault 2007, 1; cf. Foucault 2008) – *vita nuda* is the result of precisely this reduction of life to biopolitics (chiefly Agamben 1998). A state of exception such as that of the 'Greek crisis,' to which I will turn later, is so designed as to reduce human lives to the bare semblances of themselves – mere breathing bodies only kept alive so that they be held under strict surveillance and constant threat of punishment (Agamben 2005, 1–31; cf. Butler 2004). Classicizing imageries – and the example of the Archaeological Society discussed here is just one of many – are then deployed as a biopolitical apparatus in order to implement a political as well as social state of exception facilitated by the crisis and its accompanying rhetoric, thus effecting the transformation of Athens into a neoliberal dystopia.

As a social and economic system promoting austerity, deregulation, privatization, and the end of the welfare state (e.g., Harvey 2007, 64–86), Neoliberalism deploys its biopolitics as social mechanics with unprecedented force. Biopolitics plays indeed a key role in effecting the resilience of Neoliberalism, if not its thriving 'under crisis' (Mavelli 2016), and the Greek sovereign debt crisis is a case in point: in 2010 Greece was placed under the joint custody of the European Union (EU), the ECB (European Central Bank), and the IMF (International Monetary Fund), following the country's inability to sustain its sovereign debt. Bailout loans were obtained in 2010, 2012, and 2015 to that effect. In 2015 Greece became the first 'developed' country to fail making an IMF loan repayment, which led to a several-weeks long bank closure, with national debt reaching €323 billion (€30,000 per capita). Unemployment had climbed to an unprecedented 28% in September 2013 (and unemployment for 18–24-year-olds was calculated at 60.8% in February of that year); within a space of five years, the percentage of the country's population on the verge of poverty or social exclusion increased from 28.1% in 2008 to 35.7% in 2013. In that same year, over 44% of the Greek population was calculated with an income below the poverty line resulting from a series of strict austerity laws (See Eurostat 2017; General Confederation of Greek Workers 2013; Hellenic League for Human Rights 2014; and cf. Michael-Matsas 2013). As reported by The Committee on Civil Liberties, Justice, and Home Affairs of the European Parliament, austerity politics in Greece resulted in severe inequality, with special emphasis on public education, public health, the right to work, and access to justice (LIBE 2015).[7]

In this chapter I am considering examples of such Greek biopolitics in the framework of the Greek crisis and its aftermath against certain non-canonical, often performative responses to Neoliberalism and its politics of precarity (cf. Douzinas 2013, 32–48). In this, I will be examining how the display of corporeal vulnerability can become a social strategy, at the moment when, in the words of Judith Butler, 'the uncounted prove to be reflexive and start to count themselves, not only enumerating who they are, but "appearing" in some way, exercising in that way a "right" (extralegal, to be sure) to existence' (Butler and Athanasiou 2013, 101). My main case study on this end will be the city's extensive street-art scene, involving a considerable number of graffiti artists, muralists and writers, providing it would seem the landscape of the Greek crisis with the appropriate scenery in which to unfold (see, most recently, Tulke 2017, 2019).

For what the Archaeological Society at Athens is denying to the city's homeless, is generously offered – albeit rather metaphorically – to them by this more recent form of art, one most likely failing the aesthetic criteria of the Greek elites on all counts: in the streets of Exarcheia, a downtown Athenian neighbourhood only a few streets away from Omirou Street, we find, by way of a famous example, a homeless man comfortably resting, splayed along the walls of a derelict neoclassical building (Figure 9.2). The creator of this mural is the internationally known Balinese street artist WD (short for 'Wild Drawing'), currently based in Athens. Although he has worked in many of the world's capital cities, WD's murals have come in recent years to represent Athens and its multi-cultural communities, severely afflicted by the tortuously long economic recession.[8] His dystopic creations, where the unexpected strength of neo-representational melancholy counteracts the deformed face of urban frustration in a city – and a society – in the verge of collapse, are often taken to provide forceful, as well as meaningful imageries of the biopolitical terror generated by the collapse of the Greek welfare state.

WD's work, as well as that of several other street artists operating in Athens in recent years, attempts to generate a counter-environment within an urban dystopia of unemployment and homelessness. The deformed faces of his protagonists pierce the urban landscape in their silent frustration, bewildered by their own agony. They often cast an ironic gaze on symbols of capitalist prosperity, such as the five-euro note, reconstructed on the wall of a half-ruined neoclassical house in Exarcheia in 2014 (Figure 9.3), at a time when rumours of a Greek withdrawal from the eurozone (mockingly dubbed 'Grexit' by EU bureaucrats) were instilling an unprecedented sentiment of financial insecurity across Greek society and enabled the enforcement of even stricter austerity laws.

Figure 9.2 Athens, Exarcheia; 'No Land for the Poor'; mural by street artist WD (2015).
© Photo and source: Dimitris Plantzos (2017).

Figure 9.3 Athens, Exarcheia; mural by street artist WD (2014).
© Photo and source: Dimitris Plantzos (2017).

In much of his other work, however, WD seems to appreciate the city's classical and neoclassical past to the point of idolizing it. In his famous 'Knowledge Speaks – Wisdom Listens' street mural from 2016 (Wild Drawing 2019), the artist uses the corner of a humble, and now uninhabited 'neoclassical' ground-floor house in the Athenian neighbourhood of Metaxourgeio in order on the one hand to convey an esoteric (though endearingly banal) message to his viewers and on the other to express his melancholic nostalgia for the city's classicizing past, which he is in the process of reimagining through his work.[9]

In one of his less-known works, showing a kneeling Caryatid consulting her mobile phone, WD develops further his pessimism toward modernization (and modernity itself), turning nostalgically to a place and time of classical tranquillity (Figure 9.4). The work, titled 'Break Time,' adorns the side wall of a warehouse in the Athenian area of Votanikos. Formerly a working-class district with modest houses and clusters of paint shops, garages, and automobile repair yards, the neighbourhood is currently undergoing a swift gentrification process under the up-market pressures of loft-housing and Airbnb rentals, an influx of drinking and dining establishments mostly catering for the city's post-hipster subculture, and a booming underground theatre scene. The mural (indicatively facing an old car-paint workshop now turned to a theatre called *Vafeio*, which stands for 'paint workshop' in Greek), shows a classical Greek female statue at ease, taking a break as it were from her arduous task of carrying the weight of a classical temple next to her sisters. This is what the Greeks called a 'Caryatid' and what the Roman elites adopted as a decorative element of their own architecture, before it was finally taken up by the architects of the Renaissance and the Greek revival. In Modern Greece, owing to a great extent to the abduction of one such statue from the 5th-century Erechtheion temple on the Acropolis by Lord Elgin in 1809, Caryatids have come to symbolize the diachrony of Greek culture at large – in part seen as the nation's mothers and in part as its long-lost daughters (Plantzos 2017).

All this is still evident in WD's piece: the girl is shown in a convincingly rendered, Greek-like garment, her gentle profile accentuated in the style we call 'classical,' though a bit exaggerated, so as to look slightly foreign (Asian or African perhaps); this racialization of the otherwise classical face generates a sense of ambivalence to the viewer – is this a 'Black Caryatid' subtly undermining the ages-long narrative of classical culture as a 'superior,' racially defined, accomplishment? WD does not seem to be pushing this idea too far, and the title he has given to his work suggests otherwise: as Caryatids were in fact pillars carrying the building's entablature (what the Greeks called the 'geison'), they needed a capital in mediation between themselves and the load of the architrave they supported. In the Erechtheion and other classical temples this took the form of a ceremonial basket 'worn' by the maidens as a monumental cap of sorts. WD develops this ancient metaphor to its extreme, showing his Caryatid bare-headed, with her 'cap' (complete with its Ionic egg-and-dart moulding and the rectangular abacus it was used to support) casually resting on a three-dimensionally rendered mantel next to her. Accentuated by the meticulous *trompe l'oeil* WD likes to use, the piece's classicizing elements (as is also the case with the 'Knowledge-Wisdom' mural mentioned previously) deploy a neo-representational visual breed of art both in

Figure 9.4 Athens, Votanikos; 'Break Time'; mural by street artist WD (2015).

© Photo and source: Dimitris Plantzos (2016).

an attempt to mobilize a general viewing public conditioned to accept (neo)classicism as its own cultural genealogy, and as a weapon against the counter-cultures generated by the crisis. What WD seems to be arguing with these pieces is that Greece's modernization has effected the loss of its classical soul in favour of the

misleading lures of westernization (here represented by the mobile phone attracting the attention of his Caryatid). To a great extent, this is a conservative proposi-
tion, introvert and technophobic, and framed as a 'back to basics' approach to a
state of exception.

In certain ways, therefore, WD's art appears to be deploying a biopolitics of
its own; shaken by the violence of the crisis and the strict austerity implemented
in its aftermath, he draws on Greece's classical past in pretty much the way the
Greek state itself, or the various established authorities supporting the current
regime, rely on the nation's cultural capital in order to enforce their own rule.
As I have argued elsewhere, cultural symbols entangled in social wars over identity are endowed with the inherent potential for multiple signification (Plantzos
2012, 229–237). In the following section, I will consider street art as such a sort
of 'grassroots biopolitics' in the wider framework of the Greek crisis and its
aftermath.

Sovereign power and bare life

As an unauthorized medium of public expression allowing minority, nonhegem-
onic voices to be heard, graffiti and street art in general have acquired noticeable
momentum in Athens in the austerity years (Tulke 2017). A counter-discourse is
thus emerging, employing an unsolicited, violent even, new breed of visual culture challenging the dominant media representations of Greece and the Greeks at
home and abroad (Karathanasis 2014).

Nevertheless, graffiti art is not new in Athens, nor is it exclusively tied to the
'crisis,' however one may choose to define it. Already in the late 1990s, an older
generation of street artists and writers had tried to express their frustrations and
anger against a hegemonic culture of urban modernization and civic order, mockingly confronting those who believed that Athens, and the country at large, had
successfully managed to overcome the challenges posed by late modernity and
globalization. Raised in a crude consumerist culture that was soon to prove unsustainable, and disenfranchised by the equally unsustainable economics of an ever-
expanding capitalism that seemed to invite everyone yet include no one, Athenian
street artists of the 2000s embarked on a 'culture-jamming' or 'culture-hacking'
mission they imported from elsewhere (see Lewisohn 2008, 115–117). Just when
Banksy and other such dissident street artists in Great Britain and the US were
entering the scene, a bunch of Athenian artists were making themselves known
with their own 'subvertising' campaigns (as opposed to the massive, glamourous
advertising projects Greek cities were exposed to at the time).

Street artists styling themselves as Pete, Dreyk the pirate, Joad, b., and Quits –
to name but a few – turned to the city, and the bodies living in it, in an effort to
express a political message of disquiet and disillusionment (anger came later).[10]
Their dark, often morose work (especially Pete's, cf. Plantzos 2008b) seems in
anticipation of the anguish of the austerity years; they sometimes come across
as aggressively activist, though more often they seem to promote a newly found

romanticism, which was to dominate street art of the 2000s and the 2010s (mostly Joad, cf. Flickr 2007). These artists, in their early to mid-twenties at the time, concentrated their work in the cracks of the urban landscape of an emerging late-modern metropolis, an Athens that had long ceased to be called 'a village.' As the city was re-inventing itself though the intensive modernization of a newly-built underground railway, a fancy (though largely ineffective) tramline, a brand-new airport and a new motorway crossing Attica from east to west, and as the post-Olympic blues were finally kicking in after the 2004 Games, the biopolitics of gentrification were about to be employed in Athens, its impoverished downtown, and many of its hitherto neglected suburbs.

The work those street artists produced spoke of the biopolitics of gentrification affecting entire neighbourhoods in the Athenian downtown and its peripheries at the time – Gazi, Kerameikos, Rouf, Theseio. Through their designs, they wished to reinstate the(ir) body in the heart of a metropolis in the process of becoming its own Other. The bodies featured in those early works appear as the alienated, aggressively unfamiliar bodies of national discontinuity, the noncanonical bodies of sexual diversion, the vulnerable, unwanted (therefore invisible) bodies loitering the edges of late modernity. These essays in corporeal vulnerability chart the body as a battlefield, a body 'that is never simply our own' (cf. Butler and Athanasiou 2013, 98). Their performances of corporeal exposure – counter-sexual and counter-aesthetic at the same time – questioned the right to own and inhabit public space, claiming a right to visibility hitherto denied to them. Female street artists, in particular, often turned the feminist struggle for corporeal self-determination into a fight for visibility and the 'right to the city' (as in Harvey 2013). I still remember how, in one of my Sunday afternoon walks in the Kolokynthou area in late September 2007, I came across a visually stunning image of a girl in a bright yellow leather suit clinging on her body, thus exposing it, as if like naked, to the gaze of the onlooker.[11] Created by Joad in collaboration with b., the image had already attracted the reaction of a passer-by, who seemed appalled by the aggressive display of stark nakedness (one of Joad's main features): exercising their own right to public space, the anonymous writer (a man?) defaced the colourful figure with the improvised slogan 'immodest clothing feeds corruption, Greeks – Brothers' (ΤΟ ΑΣΕΜΝΟ | ΝΤΥΣΙΜΟ | ΤΡΕΦΕΙ ΤΗ | ΔΙΑΦΘΟΡΑ | ΕΛΛΗΝΕΣ | ΑΔΕΡΦΙΑ).

This exchange anticipates, I would think, the conflicts charted by Butler and Athanasiou (2013) in their joint exploration of urban wars of more recent years and the role of corporeal performativity in them. Whereas the artists of the original work 'fight for the right to be and to matter corporeally' (2013, 98 [Athanasiou]), the writer replying to them fights for a bodily integrity of a different sort, one more acceptable to the neoconservative aesthetics promoted by corporate capitalism and neoliberal politics (even though the slogan's backwardness may come across as counter-rational and third-worldist).

In the years to come, and especially after the 2010 bailout, displays of vulnerability came to express the disenfranchisement and mournful anger of those members of that earlier generation who had not retired in the meantime (e.g., Dreyk), as well as of a younger breed on newcomers, including migrants such as WD. His Jobless Clown, for example, also from Exarcheia,[12] is shown as a victim of the recession, a body marginalized in a world of corporate ventures indifferent to the long-outdated arts of makeshift theatre or simply making people laugh. As a street artist portraying another street artist in distress, WD comments on the crisis and its discourse as an apparatus of biopower (cf. Roitman 2014a, 1–22). The materiality of the crisis itself, the austerity regime we all are made to live (and die) in, is once again explored through neo-representational melancholy and the implicit aggression of a newly-constructed performatics of defeat. Images like these, however, scattered as they are to be seen in the dystopic late-capitalist landscapes of a decaying metropolis, manage to re-politicize the body as well as the crisis discourse itself.

At the same time, the anonymous writers of Athens' neglected neighbourhoods seem to attack urban decency and order through the construction of their own 'position of otherness' (Alexandrakis 2016, 289). Their interventions challenge common aesthetics as well as the city's own liveability: their work – more often castigated as mere 'vandalism' – underlines the perils of austerity and the need to resist to the neoliberalization of Greek society. For the canonical Others, however, this is the unnecessary rebellion of a wasted, and deplorably non-political youth. Every now and then, state and municipal authorities, fully supported by main-stream media, launch highly-publicized campaigns against graffiti and street art at large (which they invariably perceive as 'littering') that inevitably usher further gentrification plans. A good example of the biopolitical rhetoric usually deployed against street art (one of many) is an editorial from Greek daily *Kathimerini* (a strongly conservative paper, habitually promoting the ideology and interests of the Greek Right) from September 2016:[13] 'We are even experiencing the absolutely crazy phenomenon of dirt being elevated to ideology' the author is complaining, cynically associating those protesting against austerity measures with some kind of pollution that needs to be cleansed for the benefit of 'our' cities. In many such reports or opinion pieces, as well as statements on behalf of the authorities, the city's centre is often portrayed as 'deserted,'[14] clearly because the thousands of homeless men and women peopling it, the undocumented immigrants, the prosti-tutes, the drug addicts, are not allowed to count as 'people.'

In contrast with these successive generations of street artists I have discussed so far, living and working in the margins of the urban tissue, a newly emerging class of street muralists seem more prepared to disseminate a hegemonic discourse on public aesthetics and at the same time be consumed by it. Artist INO (2019) is a good example of this: although posing as a street artist like any other (e.g., avoid-ing to show his face while at work, as if he is working clandestinely),[15] he mostly works on commission from state and municipal authorities or corporate clients (from the Municipality of Athens to the Parliament of the Republic of Cyprus).[16] Much of INO's output refers to the current economic and social climate (from social exclusion and public surveillance to austerity and its aftermath), though a discernible strain of his work turns to classical antiquity, which he employs as a paradigm of cultural and aesthetic values now long gone. It is with these latter works that one finds that the artist has fully internalized the neoliberal, moralistic narratives regarding the Greek crisis – as a teleology of sorts, an authoritarian moralistic tale in which austerity and its forceful social impact are but a well-deserved punishment of a failing state (indeed a failing nation; see Liakos 2013; Douzinas 2013, 96–100; cf. Roitman 2014b, 22–31).

INO's mural now adorning the Cypriot parliament in Nicosia is a good example of this:[17] executed in 2016, the work (titled 'Ignorance is Bliss,' and celebrated as 'the first large-scale mural in the world painted inside a parliament building') shows (ancient) Greek political figures Solon and Pericles with their eyes covered 'so they cannot see how their ideas are used by some governments to serve specific interests.'[18] Although Solon and Pericles were Greek rather than Cypriot, and even though the Republic of Cyprus is supposedly a bilingual, bicultural state (consisting of both Greek- and Turkish-Cypriots), the Parliament at Nicosia is choosing to enforce Greek archaeolatry onto its own citizens as an exercize in cultural bio-power, as well as promote a severely (and bizarrely) anti-parliamentarian message, one usually associated with alt-right, anti-establishment populism.

A second mural by INO, this one adorning the side wall of a deserted condominium in the Metaxourgeio area in Athens, is a case in point regarding his use of classical antiquity in the framework of the Greek crisis. The work, executed in 2013, features Solon once again in the artist's trademark neo-representational style;[19] its somewhat incomprehensible title ('System of a Fraud') seems to be referring once again to the abuse of classical patrimony by the Modern Greek state and its people. As the artist stated in an interview, his composition 'foregrounds the great distance that we, as a people, have from the Ancient Greeks.'[20] Besides the work's overtly moralistic tone, however, a more specific political agenda seems to be lurking here: Solon (c. 638–558 BC) was an Athenian legislator who famously introduced a set of laws in order to relieve his fellow-citizens from accumulating debt, leading to serfdom and slavery – what the Greeks knew as *seisachtheia*. Interestingly enough, the ancient word became an anti-austerity slogan in the massive rallies protesting against the government's fiscal policies and harsh measures, especially in the first years of the crisis (between 2010 and 2012; cf. Douzinas 2013, 122–123).

INO's mural, therefore, deploys classical antiquity as a biopolitical apparatus in order to organize his audience as a congregation of archaeolatric bodies, shamed for their neglect of a classical inheritance that was not really their own, and at the same time appropriates the crisis as 'an observation that produces meaning' (see Roitman 2014b, 42). What meaning though? INO's mural is housed on a now derelict building of the Modernist era, in a neighbourhood that has most likely seen better days. *Not* by coincidence, the mural's location is embarrassingly near the premises of one of Athens's best-known art galleries, which has chosen this part of town, and a street of down-market brothels, shabby bars and squalid coffeeshops, in order to sell art to the city's elites – provided they take the risk of the adventurous trip downtown from their secured suburbs. In this way, INO's mural (which could not have been executed without expensive machinery and equipment) manages both to gentrify the neighbourhood in an ostensibly alternative and 'hip' way (by colonizing an inherently illicit art) and at the same time reinforce the biopolitical narratives of the crisis onto the bodies of its very victims. Street art thus turns into corporate merchandize (thinly) disguising itself as 'disobedience.'

It is such relics of a dispossessed (classical) heritage that INO, and other artists in his league, feel compelled to use. They are supposed to constitute a new sort of recycled Greekness, one more palatable than the globalized angst of street art in the margins (one that the mass public, let alone municipal and state authorities, would be more likely to approve of). Yet, traces of Greece's classical culture also survive in Athenian street art, most often as ironic juxtaposition with the pressing realities of the present (cf. Tulke 2017, 212, Figure 9.4). In the paste-ups of street-artist Hope, to cite the most characteristic example (Figure 9.5), imageries directly imported from classical art are projected against the urban tissue not as didactic references but as silent (silenced?) voices of dissent: working in downtown areas such as Metaxourgeio, Hope's art interferes with the marginalized world of low-market prostitution, catering for the dispossessed, excluded and vulnerable bodies of the city's unwanted masses of nameless, undocumented migrants. His subtly ironic commentaries, pasted on the relics of Greece's neoclassical dream, undermine urban normativity and propriety, at the same time foregrounding the 'real' reality of a neglected neighbourhood in the centre of Athens.

Conclusion: precarious life

'To claim the right to the city is, in effect, to claim a right to something that no longer exists,' states David Harvey, echoing Lefebvre, in his *Rebel Cities* (2013, xv). In austerity-stricken Athens, crisis narratives appear both located in culture as well as being the stuff of culture itself; for, in the words of Janet Roitman, 'crisis partakes of a metaphysics of history' (2014b, 93). The bodies generated by the Greek crisis – excluded, precarious, and vulnerable – are biopolitically reproduced as citizens and subjects only to the extent they legitimize the crisis itself, typically as justification for its enforcement in the first place. Whereas street artists of the 2000s, and even WD with his near-star status and mainstream media exposure, exploit the traces of a ruined city in order to reveal the cracks of our lives in it, INO seems to have embarked on a normalization and legitimization project, sponsored by those who wish to capitalize on slum tourism and the aesthetics of crisis in the first place. Through the work of the former, Athens emerges as a 'make-believe,' 'affective space' suffering from acute 'spatial melancholia' (Navaro-Yashin 2012, 161–175); INO's art, also reorganizing the city as an affective space, proves that heritage, like identity, does not consist of a truth essential in itself but a recurring performance rooted in the imagined materialities of culture.

Figure 9.5 Athens, Kerameikos; paste-up by street artist Hope.

© Photo and source: Dimitris Plantzos (2017).

Notes

1 This statement may be found only on the Greek-language version page of the Society; the Society's fully-functional English-language page was under construction at the time of writing. Further information on the Society, its policies, and its strategies may be sought at: https://goo.gl/ARxvGC (Accessed 10.01.2019).

2 By 'Greek lands' the text refers to the territories still (co-)inhabited by populations identifying themselves as Hellenic, which at the time included parts of the Ottoman Empire such as Thessaly, Epirus, Macedonia, and Thrace (now incorporated to the Greek State in whole or in part) as well as areas in what in Antiquity was Asia Minor and now form part of modern Turkey. On the narratives and delusions associated with the irredentist dreams of Greek intellectuals throughout the 19th and early 20th centuries, see briefly Clogg 1992, 46–97.

3 See https://goo.gl/NVkC6a (Accessed 10.01.2019).

4 On archaeology and the Greek state in the 19th century, see Herzfeld (1986, 10–11); Peckham (2001, 115–136); Gourgouris (1996, 145–148); Hamilakis (2007, 57–123).

5 For a survey of the transformation of Athens through strategic city planning, see chiefly Bastéa 1999, 69–104. See also Biris and Kardamitsi-Adami (2001, 69–102); Philippidis (2007, 44–71).

6 Homelessness in post-capitalist societies is a multi-faceted phenomenon, comprising a wide array of habitation regimes, visible and invisible. If one counts cases of insecure or unfit housing as a form of homelessness, then the number of homeless men and women in Athens and the wider Attica area (inhabited by 45% of the entire Greek population) rises up to 15,000 (with an estimate of up to 21,000 for the country as a whole); a survey conducted in 2015 by the European Federation of National Organisations Working with the Homeless concluded that approximately 17,000 people were sleeping in the rough in the prefecture of Attica during the winter of 2014–15 (FEANTSA 2017).

7 On the Greek crisis and its aftermath, see chiefly Douzinas (2013); Liakos (2013); Rakopoulos (2014); Tziovas (2017); Dalakoglou and Agelopoulos (2017).

8 WD's first engagement with the street art scene in Athens apparently dates from the 'December' riots of 2008, see https://goo.gl/8a5qL7 (Accessed 10.01.2019). On 'acts of physical violence performed publicly in Athens' during the recession years as 'a social event and . . . a means of communication,' see Dalakoglou (2011).

9 During the late 19th and the early 20th centuries, working-class houses in Athens and other Greek towns habitually copied the morphology of neoclassical public buildings and mansions erected in Athens, in an attempt to 'modernize' the Greek state. Though not neoclassical as such, these humble imitations in the post-war years came to be treated as 'traditional,' encapsulating what was seen as a contemporary Greek aesthetics, drawing from the classical past and its global reception since the Renaissance; see Biris and Kardamitsi-Adami (2001, 29–50); on the development of the Greek neoclassical town, see Philippidis (2007, 28–41).

10 I am not aware of a systematic study of these artists and their peers; nor am I conducting one here. In the section that follows I chiefly rely on my own, pretty much random though quite representative, selection of photos from the city center of Athens, as uploaded in flickr.com and the blog (pre)texts I was keeping at the time.

11 See https://goo.gl/xpHsX8 (Accessed 10.02.2019).

12 See https://goo.gl/8Rn6B5 (Accessed 10.02.2019).

13 See https://goo.gl/yE8UrR (Accessed 10.02.2019).

14 Cf. https://goo.gl/pSQ2wY; https://goo.gl/zb8ZRR (Accessed, both, 10.02.2019).

15 See, for example, this short video clip from his Instagram account, where he is covered in a hood and other ghetto-style gear, even though he is working from a crane, using expensive scaffolding, etc.: https://goo.gl/FHEaNK (Accessed 10.02.2019).

16 On the effort by central authorities to use street art in order to 'beautify' central parts of Athens through the commission of massive, centrally located murals, see https://goo.gl/bpTGE7 (Accessed 10.02.2019).

17 See https://goo.gl/rXj9t6 (Accessed 10.02.2019).

18 See, e.g., https://goo.gl/wYbRTc (Accessed 10.02.2019).

19 See https://goo.gl/UYzzeC (Accessed 10.02.2019).

20 https://goo.gl/Cs2Qnb (Accessed 10.02.2019).

Bibliography

Agamben, G. 1998. *Homo Sacer. Sovereign Life and Bare Life.* Stanford: Stanford University Press.

Agamben, G. 2005. *State of Exception.* Chicago: The University of Chicago Press.

Alexandrakis, O. 2016. 'Incidental Activism: Graffiti and Political Possibility in Athens, Greece.' In *Cultural Anthropology* 31(2), 272–296.

Arapoglou, V. and Gounis, K. 2015. *In the Shadow of the Welfare State Crisis: Visible and Invisible Homelessness in Athens in 2013.* Available at: https://goo.gl/J7bX5Y (Accessed 10.01.2019).

Archaeological Society. 2019. *The Archaeological Society at Athens.* Available at: https://goo.gl/inAFkG (Accessed 10.01.2019).

Bastéa, E. 1999. *The Creation of Modern Athens. Planning the Myth.* Cambridge: Cambridge University Press.

Biris, M. and Kardamitsi-Adami, M. 2001. *Neoclassical Architecture in Greece.* Athens: Melissa.

Brown, W. 2015. *Undoing the Demos. Neoliberalism's Stealth Revolution.* New York: Zone Books.

Butler, J. 2004. *Precarious Life. The Powers of Mourning and Violence.* London and New York: Verso.

Butler, J. and Athanasiou, A. 2013. *Dispossession. The Performative in the Political.* Cambridge: Polity Press.

Clark, T.J. 1984. *The Painting of Modern Life. Paris in the Art of Monet and his Followers.* Princeton: Princeton University Press.

Clogg, R. 1992. *A Concise History of Greece,* 2nd ed. Cambridge: Cambridge University Press.

Dalakoglou, D. 2011. 'The Irregularities of Violence in Athens.' In *Society for Cultural Anthropology website.* Available at: https://goo.gl/ynRv6g (Accessed 10.01.2019).

Dalakoglou, D. and Agelopoulos, G. eds. 2017. *Critical Times in Greece. Anthropological Engagements with the Crisis.* London: Routledge.

de Moncan, P. 2012. *Le Paris d'Haussmann.* Paris: Les Editions du Mécène.

Douzinas, C. 2013. *Philosophy and Resistance in the Crisis. Greece and the Future of Europe.* Oxford: Polity Press.

Eurostat. 2017. *Unemployment Statistics.* Available at: http://goo.gl/LwkFwt. (Accessed 03.10.2018).

FEANTSA. 2017. *Homelessness in Greece.* Available at: https://goo.gl/d8tMFF. (Accessed 10.01.2019).

Flickr. 2007. *Street Art.* Available at: https://goo.gl/ksAnNV (Accessed 10.02.2019).

Foucault, M. 2004. *Society Must be Defended.* London: Penguin Books.

Foucault, M. 2007. *Security, Territory, Population.* New York: Palgrave Macmillan.

Foucault, M. 2008. *The Birth of Biopolitics. Lectures at the Collège de France 1978–1979.* New York: Picador.

General Confederation of Greek Workers. 2013. *Employment and Unemployment in Greece for 2012.* Available at: http://goo.gl/7xRrQy (Accessed 03.10.2018).

Gourgouris, S. 1996. *Dream Nation. Enlightenment, Colonization and the Institution of Modern Greece.* Stanford: Stanford University Press.

Hamilakis, Y. 2007. *The Nation and its Ruins: Antiquity, Archaeology, and National Imagination in Greece.* Oxford: Oxford University Press.

Harvey, D. 2007. *A Brief History of Neoliberalism.* Oxford: Oxford University Press.

Harvey, D. 2013. *Rebel Cities. From the Right of the City to the Urban Revolution.* London and New York: Verso.

Hellenic League for Human Rights. 2014. *Downgrading Rights: The Cost of Austerity in Greece.* Available at: http://goo.gl/KyLc7N (Accessed 03.10.2018).

Herzfeld, M. 1986. *Ours Once More. Folklore, Ideology, and the Making of Modern Greece.* New York: Pella.

INO. 2019. *Visual Artist.* Available at: www.ino.net/ (Accessed 10.02.2019).

Karathanasis, P. 2014. 'Re-image-ing and Re-imagining the City: Overpainted Landscapes of Central Athens.' In Tsilimpounidi, M. and Walsh, A. eds., *Re-mapping 'Crisis.' A Guide to Athens.* Ropley: Zero Books, 177–182.

Korbiel, I. and Sarikakis, K. 2018. 'Media and Citizens in Greece and Beyond: Resistance and Domination through the Eurocrisis.' In *Journal of Greek Media and Culture* 4(1), 91–106.

Lewisohn, C. 2008. *Street Art. The Graffiti Revolution.* London: Tate Publishing.

Liakos, A. 2013. 'Greek Narratives of Crisis.' In *Humaniora. Czasopismo Internetowe* 3(3), 79–86. Available at: http://goo.gl/hqILcL (Accessed 03.10.2018).

LIBE. 2015. *The Impact of the Crisis on Fundamental Rights Across Member States of the EU. Country Report on Greece.* Study for the LIBE Committee. Available at: http://goo.gl/2jFRkj (Accessed 10.01.2019).

Mavelli, L. 2016. 'Governing the Resilience of Neoliberalism through Biopolitics.' In *European Journal of International Relations* 23(3), 489–512.

Michael-Matsas, S. 2013. 'Greece at the Boiling Point.' In *Critique: Journal of Socialist Theory* 41(3), 437–443.

National Hellenic Research Foundation. 2019. *Contemporary Monuments Database.* Available at: https://goo.gl/q4hsW5 (Accessed 10.01.2019).

Navaro-Yashin, Y. 2012. *The Make-Believe Space. Affective Geography in a Postwar Polity.* Durham: Duke University Press.

Peckham, R.S. 2001. *National Histories, Natural States. Nationalism and the Politics of Place in Greece.* London: I.B. Tauris.

Philippidis, D. 2007. *Neoclassical Towns in Greece (1830–1920).* Athens: Melissa.

Plantzos, D. 2008a. 'Archaeology and Hellenic identity, 1896–2004: The Frustrated Vision.' In Damaskos, D. and Plantzos, D. eds., *A Singular Antiquity. Archaeology and Hellenic Identity in 20th-Century Greece.* Athens: Benaki, 10–30.

Plantzos, D. 2008b. 'Steet Art' (Pre)texts. (Updated October 2014). Available at: https://goo.gl/i8iEnc (Accessed 10.02.2019).

Plantzos, D. 2012. 'The Kouros of Keratea. Constructing Subaltern Pasts in Contemporary Greece.' In *Journal of Social Archaeology* 12(2), 220–244.

Plantzos, D. 2017. 'Caryatids Lost and Regained: Rebranding the Classical Body in Contemporary Greece.' In *Journal of Greek Media & Culture* 3(1), 3–29.

Rakopoulos, T. 2014. 'The Crisis Seen from Below, Within, and Against: From Solidarity Economy to Food Distribution Cooperatives in Greece.' In *Dialectical Anthropology* 38(2), 189–207.

Roitman, J. 2014a. *Fiscal Disobedience. An Anthropology of Economic Regulation in Central Africa.* Princeton: Princeton University Press.

Roitman, J. 2014b. *Anti-Crisis.* Durham and London: Duke University Press.

Tulke, J. 2017. 'Visual Encounters with Crisis and Austerity: Reflections on the Cultural Politics of Street art in Contemporary Athens.' In Tziovas, D. ed. 2017, 201–219.

Tulke, J. 2019. *Aesthetics of Crisis*. Available at: https://goo.gl/7tStpg (Accessed 10.01.2019).

Tziovas, D. ed. 2017. *Greece in Crisis. The Cultural Politics of Austerity*. London and New York: I.B. Tauris.

Wild Drawing. 2019. *WDstreetart*. Available at: https://goo.gl/T8FcNo (Accessed 10.01.2019).

Part 4

(Re)-constituting the sublime image of the city

Nature, architecture, and politics in art representations of modern and ultramodern space

10 Urban gardening as a collective participatory art

Landscape and political qualities related to the concept of the 'sublime'

Konstantinos Moraitis

Introducing landscape and garden art: is collective participatory gardening 'sub-lime,' surpassing restricted individual limits?

In Exarcheia, Athens, in a rather densely populated part of the city, an extended graffiti painting on a building façade depicts a jungle scene of tropical plantation, made during the last years of economic crisis. Behind the first denotative layer of the painting, concerning the volition of the street artist to bring his/her art within public view, a second connotative layer appears. It reveals the extended demand of the Athenian urban population for 'green urbanism,' which would insert natural plantation in the interior of the Hellenic capital. Thus we may refer, besides graffiti art, to a second art, surely more important than the previous, concerning urban formations in general; it is the art of urban gardening.

We shall present this second art as highly correlated not only to environmental but also to political needs, to the 'political sustainability' of the urban ethics. Looking backwards to European and Western history, we may comment that since the first appearance of public parks in 17th-century Holland (De Jong 1990, 27–30), urban gardening and urban parks were closely associated to the democratic ethics of Western societies, to their need for environmental amelioration, as well as to the idealized landscape paradigms of the promising Arcadias, to the Elysian fields where political and personal everyday freedom could be attained. It was in this overall political context that the 18th-century 'park movement' was expanded all over the Western world, conveying its emblematic references to the cultural landscapes of Roman and Hellenic antiquity, to the Athenian ancient urban-scape par excellence, as reminiscences of the remote origins of the neoteric democracy.

It is argued that it would be a decision of serious political involvement to activate the contemporary Athenian population toward a collective movement of 'landscape urbanism' and 'urban gardening,' founded on genuine participatory procedures. However, it would be equally important to remind to the contemporary inhabitants of the Hellenic capital city the emblematic affinity of the cultural landscape where they live, with the genesis of the neoteric democratic Western identity. In any case we shall try, in the following text, to extend the usual reference to the terms 'sublime' and 'sublimity' by correlating them to the feeling of belonging to a social, cultural, or political group; to the feeling of community and thereby to social collective principles, which we ought to respect. It is, in this very sense, that collective urban gardening may be presented as a sublime activity, insisting on the ethics of collective principles as well as on the environmental and political utopia of landscape, reintroduced into urban formations.

Landscape and garden art as a political art

Describing landscape and garden art as a primary political art from the Renaissance till the contemporary era is not mere eloquence. On the contrary, it is clear that the previous arts represent the dialectical relation of Western civilization with its opposite, namely with the natural spatial substratum It is in this overall context that landscape and garden art may depict the historically different Western cultural approaches to the evaluation of nature; they present a state of mental response that may be compared to a 'transcendental' ethical percept. It was Immanuel Kant who introduced side by side, in his *Critique of Practical Reason*, the moral law within him and the starry heavens above him, as 'the two things that fill the mind with ever new and increasing admiration and awe' (Kant 1999, 269). He thus compared the formation of social and, according to us, political principles, the moral law transcending limited personal principles, to an all-encompassing landscape perception that clearly overwhelms human scale; that of the sky dome which could be qualified as 'sublime.'

We may hence return to the title of this book which ensued from the conference entitled *Art and the City*, held in March 2017. In which way an activity as common as urban gardening could reclaim the feeling of sublime? The answer to this question has already been presented in brief, and it could be analyzed in detail as a second step of our argumentation. Firstly, however, we have to understand the way landscape is correlated to concepts producing the feeling of sublimity, not in terms of natural existence only, but also in terms of cultural formation. Even more we have also to understand the way sublimity is correlated to collective formation and to principles surpassing, 'transcending' individual existence.

Political involvement and 'sublime' conditions transcending individual existence

It is not bizarre that an important initial analysis of the idea of the sublime was written by a statesman and political theorist of the 18th century, Edmund Burke, widely regarded as the philosophical founder of modern conservatism. In his *Philosophical Enquiry Into the Origin of Our Ideas of the Sublime and Beautiful* (Burke 1823), Burke proceeds to a minute analysis of all possible conditions, in natural environment or under cultural influence, that may produce awe or even terror, which could be correlated to the feeling of sublime. We may suppose that it would be important, for a theorist insisting on political stability and preservation of religious sentiments, to understand all stimuli creating feelings of inferiority and subordination to individuals, being thus subjected to the possibility of central control. Or, using a rather positive description, it would be important for a political theorist to understand the terms that produce to individuals the sentiment of belonging to a wider social structure imposing to them a sense of respect.

In both cases, of negative or positive political appreciation concerning the sublime, the feelings associated to it are produced by conditions surpassing individual existence. Within this context, religious structures that tend to produce the previous response may use expressive means analogous to emblematic forms indicating political references. It is in this overall context that the metaphor of the sky dome as used in temples, in Byzantine churches or in the Roman Pantheon, was connoting the idea of the encompassing totality of a religious belief. An analogous cosmic metaphor of the earth globe was used in the depiction of *The cenotaph for Isaac Newton*, by Étienne-Louis Boullée; the latter being, however, regarded as an emblematic reference to the non-religious ideas of the Enlightenment, to the political desire to reform the totality of political existence at a scale comparable to the recreation and the Newtonian, rational control of the whole universe. It would be rather interesting to continue this profane comparison between religious response and political, principally atheistic expression, and refer to the ceremonial festivals, organized in France after the success of the 1789 Revolution. At first glance it is astonishing that a victorious political movement, being in contradiction to conventional piety, seems to imitate religious expression.

However approaches as the one proposed by Mona Ozouf (1976) in her book on *Festivals and the French Revolution* insist on the description of a new, post-1789 attitude toward space and time; a new spatio-temporal attitude, which had to be installed as a result of the political revolutionary change. 'A rational program of spatial occupation' had to be created, she remarks, 'as well as . . . the disaffection of churches and the destruction of the symbols of the ancient regime' had to take place. Open-air festivals with the participation of the biggest possible numbers of citizens could connote the sacralization of the newly born democratic condition, the promotion of its undisputable importance, imposed to individuals through ritualized cult. Thus we could mention not only the revolutionary festivals with symbolic reference to Reason, but also those dedicated to the Supreme Being, as if the revolutionary reorganization of society had to be succeeded by the re-establishment of the cosmic totality and the deification of the new political approach. In any case, political ideals were projected on the urban landscape at a scale much greater than individuals' natural and psychic 'dimensions,' at a scale imposing the feeling of sublime in correlation to principles, ethical principles of collective and probably universal validity.

Thus the two characteristic Kantian references that 'fill the mind with ever new and increasing admiration and awe' appear side by side: social principles and the sky dome worshiped with the 'stars' of a newly born regime. It is also significant that the previously described revolutionary festivities are characterized by historians as possessing 'carnivalesque' identity, the carnival having to do in European medieval tradition with the apocalyptic destruction of natural and social continuity, making possible the subversion of existing norms. They could thus disregard 'the structures and routines of (conventional) everyday life' (Leclerc 2019, 2), and finally produce a sublime, transcendental scale of reference, able to annihilate pre-existing principles and secure a new social order, in a quasi-natural way.

The sublime in nature as a landscape quality

During the second part of the 18th century appeared the first important appraisal of romantic ideas, insisting on the supremacy of nature, a validation of the ethical importance, highly correlated, as already stated, to the concept of the sublime. This tendency was even more accentuated some years after the French Revolution, as a reaction to the first disappointment that followed its inability to realize the revolutionary promises. For the Swiss-born painter Jean-Pierre Saint-Ours (Honour 1987, 186, 207, image 107) the superiority of the classical ideals, associated with the dreams of democracy, could not retain their integrity intact. On the contrary, their comparison to the real political post-revolutionary condition could not but cause their collapse . . . as if they were destroyed by an earthquake. The metaphor of a sublime landscape of physical destruction, in Saint-Ours' depictions, connotes the destruction of hope; the loss of belief in the creation of a democratic, new society, whose political and social virtue could be compared to the ideals of classical antiquity. We refer to the painter's production after the French Revolution; to exemplary pessimistic paintings as *The Monumental Earthquake – Le Tremblement de Terre Monumental* (1792–1799), or *The Earthquake or Allegory of Terror – Le Tremblement de Terre ou Allégorie sur la Terreur* (1806), presenting immediate connotation reference to 'The Reign of Terror,' or 'The Terror – la Terreur,' the period during the French Revolution after the establishment of the First French Republic. In an apocalyptic way, Saint-Ours' paintings associate the terrifying impression of a natural destruction to the terrifying possibility of social and political collapse.

In the art history of romantic painting, representation of terrifying natural scenery was rather common. However, landscape and nature have not necessarily to be destructive in order to ensure the character of its sublimity. Natural possibility of the re-generation of life, the vital energy innate in nature, could produce the response of admiration, the recognition that natural relations could be superior to the human intellect; they could even dictate primordial paradigms of social and political affinity, in many cases much more reliable than the corrupted 'mature' Western morality. It was in this overall flux of ideas that political progress was correlated to the demand of the liberation of the natural part of human existence, of its very body; to the demand of free erotic joyance, as free as the symbolic nudity in the idealized landscape of Arcadia. Was paradise sublime and, moreover, could landscape utopias of the post-Renaissance Western world retain the idyllic atmosphere of a new political and erotic, paradisiac landscape? It seems that Faust did not turn toward Arcadia in vein, asking to the phantom of eternal beauty, to the phantom of Helen of Troy, to escort him to the seductive landscape of central Peloponnese, where their joy Arcadian would be, and free. 'Lured here to tread this blessed ground,/You fled towards a happy

Figure 10.1 Adam and Eva in the Garden of Eden – Der Sündenfall. Albrecht Dürer (1504).
A sublime landscape, and, later the Fall of Man . . . Is Western imagery tormented
by the exclusion from its natural environment?

© Photo: Wikipedia Commons. Source: https://de.wikipedia.org/wiki/Adam_und_Eva_(Dürer).

destiny!/Let our thrones as arbours now be found,/Our joy be Arcadian, and free!' (Goethe 2003, 390, verse 9570).

American Romanticism, Thoreau, and the American Transcendentalism, were forcefully oriented toward unmediated nature, in the extended wild-scape of the new world, while Europeans were trying to reinvent wilderness next to their civilized space formations. They have turned to the 'wild garden' approaches or to a hallucinatory fairy-tale gardening, as in Alice's adventures in Wonderland. Neglecting the last enigmatic reference to a magic world where flowers spoke and animals practiced human activities, we may explain that the notion of the 'wild garden' was produced in 19th-century Britain by William Robinson and influenced gardeners and landscape architects on both sides of the Atlantic, from the likes of England's great garden designer Gertrude Jekyll to Frederick Law Olmstead, designer of some of the greatest parks in North America. 'Wild garden' is a feeling of natural existence surpassing cultural formations, fantasies invading the context of reality, creation of new symbolisms of existence.

The sublime object of ideology

Fantasies, symbolic structure, reality . . . a mixture of psychic and cultural ingredients that could also have political value and hence be described by the term 'ideology,' a term being largely used in order to denote systems of cultural approach being partly 'real' and partly fictitious. The term was initially used in a pejorative way to express conditions that becloud political conscience, creating barriers between members of a given society and the supposed 'reality.' However a more sophisticated analysis could prove that total objectivity is never possible. Reality under these terms is always perceived through a veil of fantasies and is always confronted with the symbolic order, with a nexus of meanings that ascribe partial significance to any singular member of the surrounding substance. According to this approach, intensively theorized by the French psychoanalyst Jacques Lacan, and presented through the metaphorical diagram of the 'Borromean Knot' (Evans 1996, 162–164), real conditions could not be touched without the interference of our desire, of our fantasies projected on it, or without the symbolic structure attracting meaning to prescribed semiotic directions. It seems moreover that this three-part mixture of cultural perception, imaginative–symbolic–real, may be proved extremely powerful; not because of its stability, but rather because of its continuous regeneration. Imaginative projections, fantasies, are constantly succeeding one another; structures of meanings are transformed to other structures of meanings, continuously changing our appreciation of the surrounding 'real' world. In contradiction to these dialectics of change, ideology has to present itself as inalterable and certain; it can thus persuade societies of its validity, of its power, overwhelming individual possibility of denial. It is because of this supposed unaltered powerful condition of ideology that the Slovenian philosopher Slavoj Žižek speaks about *The Sublime Object of Ideology* (Žižek 2008) in a book clearly influenced by the Lacanian approach, a book that lent its title to this section of our presentation.

Then again ideology is sublime, as it surpasses individual scale, imposing modes of social expression and social existence. In a certain way ethical principles may be described as principles of ideological identity. However why is ideology referred to, as an 'object'? We could argue that ideology has to be objectified, to offer the certainty of an unaltered object, though it is under continuous alteration, 'eroded by time.' It is this last description that was used by Bernard Cache in his analysis of the continuous topological morphogenesis as well as to the computer-aided parametric design. Cache remarked that 'objectivity is eroded by time' insisting on the transformational design process that may be used in computational design. According to the previouscitation, time sequence does not only transform the feeling of every particular object, but moreover the feeling of 'objectivity' as a stabilized unaltered condition in general (Cache 1995, 96).

Ideology as an 'object' has to acquire perceptible existence, has to be represented, depicted, and constructed. It is probably on account of this necessity that Immanuel Kant had to compare ethical, ideological in our own terms, formation with the image of the starry dome of the sky. Because of this necessity romantic painting describes sublimity through landscape scenery. It is in the same context that neoteric and contemporary societies describe their possible environmental and social regeneration through urban parks, through the correlation of 'industry with agriculture' and recently through landscape urbanism approaches. To these sublime ideological approaches of environmental and social regeneration, we shall add the proposal of the collective urban gardening.

Collective urban gardening and ethics of 'cultural' participation

Can we recognize to this participatory garden culture the qualities of sublimity? We shall retain our answer. We may however recognize in it the effort to indicate a new direction of collective urban ethics, a positive urban ideology surpassing individual isolated aspirations. This effort is evidently not qualified as sublime because of its religious orientation; however it denotes the collective decision to reform reality, to offer an urban rebirth, an urban festivity of natural and social revival in the way Parisian citizens wanted to reorganize space and time correlation after the French Revolution.

On the façade of a two-story modernistic house, in Exarcheia, in the central area of Athens, as we have already remarked in the introductory part of this text, a huge graffiti depiction is offered to the passer-by. It represents a tropical forest, full of rich vegetation and exotic birds, in the same dream-like way that was used by 'douanier' Henri Rousseau in his own naïve paintings. In both cases a fertile utopia external to conventional civilization is described, as an imaginative reference to a landscape of archetypical innocence combined with primordial hedonism. The tropical forest possesses undeniably the characteristics of sublime; carnal demand may also be characterized, in its deepest sense, by a feeling of sublimity. What we may add as the next step of landscape imaginary in Exarcheia would be the need for its actual realization in the form of an urban park, created by the citizens of the neighborhood

Figure 10.2 A tropical forest in Exarcheia, Athens.

© Photo and source: Konstantinos Moraitis.

Figure 10.3 The self-managed park of Navarinou Street in Exarcheia, Athens.

© Photo and source: Konstantinos Moraitis.

themselves. We refer to the self-managed park of Navarinou Street, which was produced after the occupation of an open-air parking lot by the citizens of the territory in 2009. On March 7, 2009, the 'Citizens' of Exarcheia Initiative Committee,' in collaboration with the local collectivity 'We, Here and Now for All of Us' 'organized a protest manifestation, demanding what they considered as self-evident; to transform the parking to a park!'[1] They removed the asphalt paving, brought plantation soil with trucks, planted trees and flowers, and celebrated. Let us return to a previous comment: did they disregard 'the structures and routines of (conventional) everyday life' promoting a new proposal of urban ethics? We could assert that such an activation of collective conscience may produce a 'sub-lime' condition, which trespasses the limit; surpasses the threshold, or the *limen* in Latin, of the individual response. We could even assert that finally, religious sentiments on the one hand, or 'faith' to political principles on the other may acquire their sublime character on the ground of the same collective force; sanctuaries and houses of parliament being the emblematic structures dedicated to it, to a supposed collective volition many times intercepted, sometimes associated to a supernatural power, sometimes corresponding to an elective approval.

In neoteric Western world, this collective volition was often correlated to public parks, being in many cases the outcome of a top-down response to social calamities, unable to cure real political contradictions. However, they were also regarded, in many cases, as the 'sanctuaries' of radical political and environmental approaches, as the bottom-up green 'emeralds' of the neoteric urban utopias. 'Emeralds' as the elements of the 'emerald necklace,' in the metaphor used by the great American landscape architect Frederick Law Olmsted, in order to describe a number of parks, correlated through 'green' parkways in the urban territory of Boston, United States (Beveridge and Rocheleau 1998, 83–89).

A critical remark, on the importance of analogous urban landscape interventions, could insist on the political inadequacy of the isolated environmental amelioration; on the fact that Hausmannian creation of the Parisian green network in the mid-19th century French capital, did not succeed in postponing the revolutionary subversion of the 1871 Commune of Paris. In comparison to this negative attitude, we may nevertheless state that garden cities retained their natural seduction while progressive thinking, in many different cases, asked for the juxtaposition of agricultural activity to industrial productivity and educational development as well. Frederick Law Olmsted described agricultural communities of high educational level as an ideal condition for future societies, whereas the prince of European anarchism, Piotr Kropotkin, proposed a close correlation of agriculture to industrial activity of balanced scale and manual work to 'brain work' in self-organized communities (Kropotkin 1901).

Fertile city and Bruno Taut's crystal utopia in the alpine landscape

During a recent exposition under the title 'La Ville Fertile, Vers Une Nature Urbaine' ('The Fertile City, Towards an Urban Nature') (Paysages Actualités

2011) at the 'Cité de l'Architecture et du Patrimonie' ('the City of Architecture and Heritage') in Paris, 2011, the interest for landscape urbanism and green urban policy was visualized through a poster representing an urban formation of crystal-like fictional buildings. However, the building structures were depicted emerging from a tropical forest, as rich in vegetation as the imaginary paintings of Douanier Rouseau, or as the graffiti on the house walls of Exarcheia. In this way, contemporary imagery correlated a fascinating natural environment with a magnificent, sublime architectural image. Nevertheless, if we return to Bruno Taut's architectural fantasies at the beginning of the 20th century, we may locate another urban utopia, comparable to 'The Fertile City.' It had to be created in the Alpine region, 'between Monte Rosa and the plains of Italy,' constructed only with glass buildings, in an equally breathtaking natural scenery. It would produce an 'extra-political . . . purely human and cosmic-religious' cultural condition, using architecture and its relation to the sublime mountainous landscape as a metaphor of a new life.

We think that in both cases, the proposal of 'a new life' is of utmost importance; a new life which, in any case, could not be external to the participation, to the active volition of the people involved. Upon this concept we may correlate Taut's and Fertile City's utopias. Those exemplary proposals may use a cosmic metaphor, the metaphor of a total transformation of the conditions of existence; however, they are not 'extra-political.' They are, on the contrary, purely political though immature in their utopic, simplistic imagery. This metaphor of the total cosmic transformation may be considered as political as were the urban landscape ceremonies of the French Revolution, expecting to attend, undoubtedly, a political and ethical sublime validity.

Western identity and the sublime political reference to the Athenian ancient landscape

We can refer additionally a third example of urban utopia being strongly political, in contradiction to the two previous examples. If for Immanuel Kant starry heaven and moral law were filling his mind with admiration, equally important for him was the maturity of the citizen, the courage to use his/her own understanding; '"Sapere Aude," have courage to use your own understanding!' That was according to him the motto of the Enlightenment, indicating democratic political formations (Kant 1992). Those political formations, anyway, possessed two definite references, ancient Rome and ancient Athens, the latter being identified with the origins of European and Western democracy. Thus we may describe the Athenian landscape as sublime, not for its natural existence principally, or for its sunny weather, but because of its monumental validity and its intangible political heritage.

We refer to monumental Parthenon, to Kerameikos cemetery, Plato's Academy, Aristotle's Lyceum, Pnyx, to Colonus area where according to the myth the blind outcast Oedipus arrived, and finally to the ancient Agora. We also refer to the value of the ancient textual heritage, to what was written and thus commemorated, represented through pictorial arts, but also to what was never depicted but was still hovering in the cultural atmosphere of the Western world, as a remote call of an ancestral civilization, identified with the neoteric and contemporary need of values. The cultural importance of all those material places and their immaterial connotations, of their cultural sublimity, of their over-the-conventional-limits cultural importance was established on their solid political foundations; on political foundations analogous to those on which contemporary collective participatory activities in the city are trying to be developed. In this context, important political principles as those of equality and democracy may be associated to immediate participatory practices, to the responsible participation of each individual to the community. Ancient venerated monuments may be correlated to the everyday 'monumentality' of public participation, to the urgent need for contemporary 'Agoras' of collective political activity.

Conclusion: the sublime landscape of democracy

We have already stated that in historical terms the 'sublime' was largely correlated to natural terms, to sublime landscape whose importance could surpass individual

Figure 10.4 The sublime landscape of democracy. *Pericles' Funeral Oration – Perikles hält die Leichenrede*, by Philipp Foltz, 1852.

© Photo: Wikipedia Commons. Source: https://en.wikipedia.org/wiki/Pericles_Funeral_Oration#/media/File:Discurso_funebre_pericles.PNG

human stature, to an-all encompassing 'sky' whose dome could be a reference to serene admiration or to terrifying intensity. We have already correlated this natural landscape with collectively structured ethical principles, whose extension also surpasses individual existence. Religious and public architecture could be associated to both directions; a divine orientation related to the divine magnitude of nature as well as social principles which individuals ought to respect and accept. We are going back to collective existence.

To conclude: it is obvious that the final target of this chapter is the deconstruction of the rudimentary ideas concerning the concept of 'sublime.' The reference to the activities of participatory urban culture was only the occasion to analyze one of the principal, in our view, components of sublimity. We refer, of course, to social entirety, to social structures and principles transcending individual scale. We may compare this transcendental social power, a power that transcends individuality, with nature; or we may try to inscribe it to formations of extended participation, to the 'sublime landscape of democracy.'

Note

1 See www.enallaktikos.gr.

Bibliography

Anonymous. *Elefthero Aftodiacheirizomeno Parko Navarinou kai Zoodochou Peges – Emeis, Edo kai Tora gia Olous Emas (Free Self-Governed Park at Navarinou and Zoodochou Peges Streets – We, Here and Now for All of Us)*. Available at: www.enal laktikos. gr/kg15el_eleythero-aytodiaxeirizoeno-parko-nayarinoy-kai-zwodoxoy-pigis-emeis-edw-kai-twra-kai-gia-oloys-emas_a64.html (Accessed 11.06.2019).
Anonymous. 2011. *La Ville Fertile, vers une Nature Urbaine*. Paris: Paysages Actualités, special edition 2011. Available at: http//www.lemoniteur.fr/article/la-ville-fertile-vers-une-nature-urbaine.596259 (Accessed 17.03.2019).
Beveridge, C. and Rocheleau, P. 1998. *Frederick Law Olmsted. Designing the American Landscape*. New York: Universe Publishing.
Burke, E. 1823. *A Philosophical Enquiry into the Origin of Our Ideas of the Sublime and Beautiful, with an Introductory Discourse Concerning Taste and Several other Additions*. London: Thomas McLean, Haymarket. Available at: www.google.com/search BurkeE.A+Philosophical+Enquiry+into+the+Origin+of+Our+Ideas+of+the+Sublime+and+Beautiful sourceid=chrome&ie=UTF-8 (Accessed 22.05.2019).
Cache, B. 1995. *Earth Moves: The Furnishing of Territories*. Cambridge, MA: The MIT Press.
De Jong, E. 1990. 'For Profit and Ornament: The Function and Meaning of Dutch Garden Art in the Period of William and Mary, 1650–1702.' In Hunt, J.D. ed., *The Dutch Garden in the 17th Century*. Washington, DC: Dumbarton Oaks Trustees for Harvard University.
Evans, D. 1996. *Eisagogiko Lexiko tes Lakanikes Psychanalyses(An Introductory Dictionary of Lacanian Psychoanalysis)*. Athens: Ellenika Grammata.
Goethe von, J.W. 2003. *Faust*. Part II – Act III. Available at: www.poetry in translation. com/Admin/Hardcopy. php#print_goethe_faust (Accessed 18.03.2019).
Honour, H. 1987. *Neo-classicism-Style and Civilisation*. Middlesex, England: Penguin.

Kant, I. 1992. *An Answer to the Question: What is Enlightenment?* Indianapolis: Hackett Publishing.

Kant, I. 1999. *Practical Philosophy*. Cambridge: Cambridge University Press.

Kropotkin, P. 1901. *Fields, Factories and Workshops or Industry Combined with Agriculture and Brain Work with Manual Work*. London: G. P. Putnam's Sons.

Leclerc, G. *French Revolutionary Festivals, Modern Architecture, and the Carnivalesque*. Available at: www.academia.edu/10067978/French_ revolutionary_festivalsModern_ Architecture_and_The_Carnivalesque (Accessed 17.03.2019).

Ozouf, M. 1976. *La Fête Révolutionnaire 1789–1799*. Paris: Gallimard. English transl. 1991. *Festivals and the French Revolution*. Cambridge, MA: Harvard University Press.

Žižek, S. 2008. *The Sublime Object of Ideology*. London: Verso.

11 'Trikoupis Refuses to Unveil Himself in Order Not to See.'

A memorial statue and national identity in early 20th-century Greece

Maria-Mirka Palioura

The aim of this chapter is to examine the relationship among politics, memory, and the formation of national identity in early 20th-century Athens from the angle of nationalistic narratives as articulated and imposed by contemporary official authorities in regard to urban art policies such as the construction of statesmen' statues in central Athenian space.

In January 1920, the unveiling of the memorial statue of Charilaos Trikoupis (Tricha 2016) became the subject of extensive coverage in the Athenian press. A patriotic, persistent, audacious, insightful individual, but also a man of true integrity and fierce supporter of the constitution, Trikoupis possessed a strong personality that rose to the challenges of his times. He distinguished himself as a politician during a troubled period for Greece (1875–1895), in years mired by political strife, bankruptcy, and populism, but also by profound institutional and constitutional reforms that changed the political and socioeconomic landscape of Greece leaving their mark even to the present day. In 20th-century interwar Greece — a fruitful but painful period of national struggle, social upheaval and stagnant liquidity — his memorial statue acted as a reminder of his reformist agenda and achievements in the latter part of the 19th century (Papademetriou and Anagnostopoulos 2017, 70).

'Trikoupis refuses to unveil himself, in order not to see' remarked a Greek Deputy who attended the statue's unveiling ceremony. This utterance was recorded by an anonymous journalist of the newspaper *Empros* (Anonymous, *Empros* 1920). The pretext for this witty comment was a bizarre incident that took place during the unveiling ceremony: the Greek flag covering the statue could not be drawn despite all efforts, so Thomopoulos, the sculptor, took the matter in his own hands, climbed up the statue, and dislodged the flag. This long-forgotten tidbit of history, a chance find during our study of the Benaki Museum-Finopoulos Archive, will act as a central point for the discussion to follow.

A memorial statue to build

Ch. Trikoupis dominated the Greek political scene during the second half of the 19th century. His presence was linked to crucial political events and a far-reaching modernization program that was to lay the path for the country's transition into the 20th century. As one of the foremost political figures in Greece, his death in Cannes, France, on March 30, 1896, marked the end of an era. Barely a month later, in April 1896, a 'Commission for the Construction of a Memorial to Charilaos Trikoupis' was created to collect the necessary funds among his political friends and supporters. The correspondence between the Commission and Leonidas Karystinakis —ancestor of E. Finopoulos and thrice elected deputy of the Andros Island with Trikoupis' party (Voule ton Hellenon 1986, 118–119) — is quite revealing as it attests to the commission's concerted efforts to bring their plan to fruition. Lists of potential donors were sent to representatives of various constituencies of the land. Local politicians were expected to promote the initiative to their respective communities. Contributions by donors, like L. Karystinakis, who gave 47 drachmas to the cause (see Figure 11.1), were sent by post to the Treasurer of the Commission Pericles Valaoritis — a pre-eminent member of Trikoupis' party — at 37 Panepistimiou Avenue in Athens.

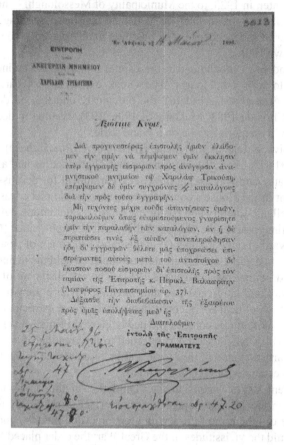

Figure 11.1 Letter to L. Karystinakis.

Source: ©The Benaki Museum – Finopoulos Collection's Archive, no. 3615.

This first attempt to set up a memorial to Trikoupis was not met with success despite the high status and influential connections of the commission's members who were close allies and friends of the deceased. Obviously they recognized the necessity of setting up a monument to generate images readily recognizable in collective imagination and constituting 'a common canvas of recording the past as a shared common past' (Stavridis 2010, 89). The undertaking quickly died out due to the humiliating defeat in the Greek-Turkish war of 1897 (Clogg 2014) and the ensuing political crisis.[1] However, the synergy of personal and collective memory (Halbwachs 1950; Confino 1997; Todorova 2004) led in November 1912, 16 years after Trikoupis' death, to the creation of a marble bust of the famous politician – funded by an undisclosed expatriate commissioner (Anonymous 1912, 177) – by the sculptor and professor at the School of Fine Arts Thomas Thomopoulos (1873–1937) (Hellenic Literary and Historical Archive, Thomopoulos Archive. Matthiopoulos 1997–2000, vol. 2, 32–33).

The bust was executed by Thomopoulos and would eventually be delivered several years later, in 1923, to the Municipality of Messolonghi – the native town of Trikoupis' family – renowned for its defiant resistance and tragic fate during the Greek War of Independence. The bust was finally placed in the environment of the Old Town Hall.

The demand for a monument in the capital honoring the eminent politician would have to wait a bit longer. The Greek expatriate Polychronis Kotsikas (1860–1922) (Tomara-Sideri 2016), the previously undisclosed donor of Trikoupis' bust (Christou 1982; Mykoniates 1996, 193), eventually decided to fund the creation of a memorial statue on a grand scale.

An extensive article in the revue *Πινακοθήκη* (*Pinakotheke*) dated in the autumn of 1914 (Δ.Ι.Κ. 1914) informs us that Thomopoulos had already built a cast of the statue and was asking the Greek Parliament for permission to set up the monument in the House's courtyard once the marble piece was completed a year later (Hellenic Literary and Historical Archive, Photographic Archives). The press of the time praised the choice of this particular spot, since 'Nowhere else but the courtyard of the Parliament —in this most prominent of positions, outside the hall which had witnessed his greatest triumphs— could the memorial statue of Trikoupis find its true place' (Anisios 1914). The establishment of a connection between history and landscape was thus attempted with a view to delineating a topos where the political order could 'temporally freeze particular values' (Verdery 1999, 6). Within this context, the newly elected Prime Minister El. Venizelos (Clogg 2014; Chatziiosef n.d.) conceded to the creation of a Commission to examine the artistic value of the statue and its resemblance to the actual person. The cost was estimated at 60,000 drachmas.

The different stages leading to the completion of the statue are thoroughly documented in the Th. Thomopoulos Archive (Hellenic Literary and Historical Archive). We can therefore follow with great precision how the undertaking was completed amid the vicissitudes of the Great War: the order placed for white marble from a Penteli[2] quarry, the purchase of other necessary materials, the wages of the workers, the transportation costs, the successive contracts with the sculptor

Loukas L. Doukas (Matthiopoulos 1997–2000, vol. 1, 387), the budget overrun, and the final expenses that amounted to 100,000 drachmas. Thomopoulos's remuneration was set at 18,500 drachmas, but even after its completion the statue was to remain for years to come in the artist's workshop (M.C. 1916) before finally finding its place in the courtyard of the Old Parliament, opposite the central staircase, overlooking Stadiou Street.

The statue and its placement

The statue is of Pentelic marble and stands mounted on an elevated pedestal reaching a total height of 6.5 meters. The front face of the plinth is carved in high relief with a winged naked youth, representing the politician's genius, sitting on the lower plinth necking. The winged figure's head is turned upward. The left arm is stretched out and raised a little above the shoulders embracing an inscription with one of Trikoupis' famous quotes: 'Greece is determined to live and will live.' At the bottom of the base, the name Trikoupis is rendered in capital letters leaving no doubt as to the statue's identification. Thomopoulos eloquently describes the monument as follows: 'I wanted to hand him down to future generations as a man of gravitas, incorruptible and full of determination: his genius sitting on the plinth and bursting with Greek beauty and strength, is inscribing that famous optimistic saying of his: Greece is determined to live and will live. . . . I did not desire to embellish the plinth with anything redundant. Trikoupis stands before the orator's podium upright, like an archaic stele, making no distasteful rhetorical gesture, with his eagle eyes and deep gaze' (Mykoniates 1996, 193).

In Thomopoulos' words we can discern his intention to invest Trikoupis with all those positive attributes that 'correspond to the ideal national body' (Galinike 2015, 344). The austere and heavy statue on which the artist displayed his great technical ability on behalf of collective imagination (Gell 2013, 218) was the second to be set up in the area surrounding the (now) Old Parliament in 1920 (see Figure 11.2). An equestrian bronze statue by L. Sochos (Matthiopoulos 1997–2000, vol. 4, 258–259) of Theodoros Kolokotronis (1770–1843), a hero of the 1821 Greek War of Independence, had already been placed in April 1904 in the homonymous square, at the junction of Stadiou and Kolokotroni streets, on the eastern side of the Parliament's building (see Figure 11.3). The decision to place Kolokotronis' equestrian statue on the islet, despite the reactions for its disproportionate size in relation to the limited space around it (Antonopoulou 2003; Daniel 2012) was taken by Sochos himself who believed that the monument would establish its own presence there.

The placement of Trikoupis' statue was to differentiate the spatial dimension of the area surrounding the Old Parliament by imposing a strong symbolism of parliamentary values. If the emergence of a Greek national state and a modern national identity after the war with the Ottomans was signified by the equestrian statue of Kolokotronis, Trikoupis' statue bridged the post-revolutionary years of the emerging Greek state with the modernizing reforms of Eleutherios Venizelos

Figure 11.2 Trikoupis' statue at its original place.

Source: © Photographic Archives, National Historical Museum, Athens.

(1864–1936), during his first tenure (14.06.1917–4.11.1920) as leader of the Liberals' Party ('*Phileleftheroi*').

Unveiling the statue

'Trikoupis refuses to unveil himself, in order not to see' remarked rather sarcastically a Greek Deputy who attended the statue's unveiling ceremony on a cold and fiercely windy winter afternoon of January 26, 1920 at approximately 12:30.[3] The witticism was recorded by an anonymous journalist of the Athenian opposition newspaper *Empros* (Anonymous, *Empros* 1920) and was reproduced with slight variations by several other newspapers (Anonymous, *Ethnos* 1920; Anonymous, *Patris* 1920). A somewhat embarrassing incident had preceded the vitriolic comment: the Greek flag covering the statue, as it was customary in such occasions, refused to give way despite the repeated and persistent efforts by the King Alexander I of Greece (1893–1920) and the Prime Minister Venizelos, forcing Thomopoulos, the sculptor, to scale the pedestal, release the obstinate flag, and give an end to this awkward episode. The deputy's fleeting remark was intended as a metaphor aimed against the contemporary political situation and the so-called *Parliament of the Lazaroi* — the unorthodox and barely constitutional 'resurrection' of the 1915 Parliament in 1917 — which had been dissolved no less unconstitutionally by the previous King of Greece Constantine I (1868–1923).

Figure 11.3 Kolokotronis' statue at its original place. Postal card.
Source: © Maria-Mirka Palioura Collection.

But yet another most interesting extract from the contemporary press tells a different story with regard to the underlying motivation behind the decision to set up the monument in this specific spot. The anonymous author stated in no uncertain terms that the Greek people should honor with statues only those politicians who respected the constitution, defended civil liberties and sacrificed their self-interests and aspirations for the betterment of their country and for the unity of the Greek people even to the point of choosing to die or go into exile (Anonymous, *Athinaiki* 1920).

In this spirit, Emmanuel Repoulis (1863–1924), the then Deputy Prime Minister, in his political memorial of Trikoupis went to great lengths to raise Venizelos to the level of the long-deceased reformist Prime Minister. The speech was read in the chamber of the Parliament that same morning before the ceremony. It linked Venizelos directly with the late Prime Minister, appropriating a part of the glory of the former and conferring it to the latter by using the telling term 'avenger' for Venizelos (Anonymous, *Aster* 1920). In this way, Repoulis tried to identify in the person of Trikoupis a political forefather and pioneer, who, like Venizelos himself, fought for constitutional and parliamentary freedom. In this vein, Venizelos the 'avenger' was promoted as the most suitable candidate to crown Trikoupis' preparatory efforts with a national triumph 'which Trikoupis was faithfully anticipating in the course of his political career' (Anonymous, *Empros* 1920). This triumph was nothing less than the expansion of the country's borders. This goal was sealed by Venizelos with the signing of the Treaty of Sèvres in August 1920 and the eventual inclusion of newly conquered territories within the Greek state. If Trikoupis was the politician who 'interweaved the external aspirations of Hellenism with reality' in the first place, Venizelos was the real instigator of 'the great achievements of Greece' (Anonymous, *Patris* 1920). In this way, the continuity of governance was normalized through its connection with the beginnings of power and a political genealogy (Fowler 2017). The bestowing of honors to Trikoupis as a precursor of the nation's great ideals had now been brought to their final fruition.

Changing the placement

In the decades to follow and until the end of the Greek civil war (1946–1949) (Clogg 2014), the fixed gaze of Trikoupis' statue would witness all major events and upheavals occurring in the city center. The same would be true for the equestrian statue of Kolokotronis who overlooked the square adjacent to the building.

The Parliament building would eventually be ceded to the Historical and Ethnological Society of Greece — founded in 1882 (Lappas 1982) — with the aim to house a museum where 'items and testimonies that illuminate the history of Modern Greece' (Historical and Ethnological Society of Greece) could be housed under the same roof. The concession was put into effect by Venizelos even before the body of the Parliament was transferred to the ad hoc remodeled Old Palace building, on Syntagma square, in 1935. For a short period the Ministry of Justice was housed in the Old Parliament Building until the permanent installation of the Museum after World War II, in 1962.

Despite the intended change of use of the building, the arrangement of the courtyard remained the same with the addition in 1931 of the bust of Theodoros Diligiannis (1824–1905), another prime minister and contemporary political opponent of Trikoupis. Even posthumously, the two men's rivalry was spatially fixed outside the very place of their political clash.

Time brought unexpected changes. The location of the sculptures and their significant role in the formation of the character of the Athenian city center in the first half of the 20th century would once more be altered after World War II and the end of the traumatic Greek civil war in 1949. In 1954, as part of the repair of the building and the redevelopment of the courtyard (Ho Attikos 1954, No 9), the bust of Diligiannis was relocated to the west side, Trikoupis's monument was moved from its original position to the eastern side (see Figure 11.4), approximately in the same spot where the equestrian statue of Kolokotronis was erected, which in its own turn was moved in front of the building's entrance occupying the most prominent position previously held by the Trikoupis Monument (Ho Attikos 1954, No 8) (see Figure 11.5).

The transposition of the two statues as part of the remodeling of the building and the surrounding area to house the Museum of the Historical and Ethnological Society of Greece was decided after the Society had rejected a governmental proposal to demolish the historic building.

The following handwritten note (30–10–1989)[4] by Vasilios Diamantopoulos, director of the Ministry of Finance and secretary of the Commission for the Study of the Remodelling and Renovation of the Old Parliament Building (1954–1957)[5] refers to a technical plan by the architect Andreas Ploumistos (1897–1962)

Figure 11.4 Trikoupis' statue at its contemporary place.

© Photo and source: Maria-Mirka Palioura.

(Fessa-Emmanuel 2009, 140–146) indicating how the area outside the building should be shaped: 'In those days, (during the discussions to demolish the building) a technical plan and a series of guidelines by the architect Andreas Ploumistos appeared in the newspaper *Estia*, indicating how the external space of the building of the Old Parliament should look:

(1) Removal of the iron railing enclosing the space on the sides running along Stadiou and Kolokotroni streets. (2) Raising to the ground the elevated garden continuous to the building on its eastern side (. . . illegible. . .) with citrus trees. (3) The statue of Charilaos Trikoupis to be transported from its position to the side of the main entrance overlooking Stadiou Street, to the

Figure 11.5 Kolocotronis's statue at its contemporary place.

© Photo and source: Maria-Mirka Palioura.

area of the eastern flank of the building that was to be leveled with by excavating and removing the raised garden. (4) In the space that would be freed by the transfer of the statue of Charilaos Trikoupis placed at the beginning of Omirou Street so as to be visible along its whole length place the equestrian statue of Theodoros Kolokotronis. (5) In the space freed from the statue of Kolokotronis open a short street joining Stadiou Street with Kolokotroni Street toward the beginning of Voulis Street.

The draft provided for the removal of the two statues, a decision which found favor with the press of the time. The removal, despite no evidence of any political intervention, should be related to the first post-civil war stable government of conservatives led by Alexandros Papagos (1883–1955), the victor of the Greek civil war and founder and leader of the *Hellinikos Synagermos* party (Koliopoulos and Veremis 2010).

Collective memory and national pursuits

In the early years of the post-World War II and post-civil war period, a concerted effort to symbolically consolidate the military victory of the National Army against the communist-led Democratic Army became increasingly apparent. The official ideological institutions and mechanisms, such as the educational system, the military or the church (Kouris 2016) attempted to justify in both symbolic and actual terms the final victory of the government forces. In this attempt, space was appropriated as a topos invested with social meaning. Monuments, being emblematic artifacts, were mustered to generate, serve, and shape a collective memory that would boost the desired national feeling through the promotion of symbols reminding and validating the survival of the nation against both external and internal enemies. The monuments, like the one of Trikoupis, became signs condensing memory related to events or persons through a conscious and deliberate abstraction. Their effectiveness was linked to their physical appearance in space 'as evidence of their truth' (Stavridis 1990, 115).

In this framework, the equestrian statue of Kolokotronis, as a powerful reference point of the 1821 Greek Struggle for Independence was far superior to that of Trikoupis. Trikoupis' statue was moved closer to the building but at the same time removed from collective memory as well during the first postwar decade. The transposition of the two statues resulted in a reversed hierarchical relation between these two personalities, altering at the same time the intensity and content of their symbolism. From 1947 onwards, the Greek Struggle of 1821 appeared ever more often in the front pages of newspapers, irrespective of their ideological affiliation, as a marker of national pride and unity through the display of works by well-known Greek artists.[6]

Against this background, and given the inextricable relationship between memory and power, the dominant collective memory imposed its choices by displaying the appropriate symbols that shaped and strengthened the preferred version of national identity. The underlying needs of the unofficial collective memory, which is oftentimes co-formed in interaction with the dominant one, become quite pronounced on a textual level in an article penned for the right-wing newspaper *Empros* which bore the title *The Two Silent Ones*, referring to the Trikoupis and Diligianis statues: 'It is better to forget and leave Trikoupis and Diligianis together, if necessary, to let them live only in the pages of History and in the memory of those who do not consider her worthless luggage.' 'They remain as always, guards of a past that faded out without leaving any memories behind.' And a little further down: 'Their place is no longer there' (Ho Attikos 1954, No 4). It is more than obvious that the author of the article denudes the statues from the memory they carried and places them firmly in the remotest 'pages of history.' This act would render them inactive limiting their role to decorative artworks that had no place in the modern era.

The two monuments respond through their relocation to the national expectations of the time. As contributors to the collective memory of the nation, according to Le Goff (1988) they comprise one of the most powerful ways for a society to record its point of view on history (Stavridis 1990, 115).

In the beginning of the 1950s Trikoupis' fixed gaze could no longer serve the desired collective references to the past. The statue no longer occupied a place that Pierre Nora calls *lieux de mémoire* (Nora 1989, 1997), that is to say, in a central place condensing the maximum amount of signifiers in a symbol of limited size.

As a result, the statue had no place in front of the Old Parliament's building any longer and therefore the official authorities, acting as agents managing the meaning of the past, decided to move the statue among the citrus trees, to the east and less prominent side of the building. But as each society 'builds its space' (Stavridis 1990, 11–12), today the emblematic character of the monument finds a new place, proving the plasticity of memory.

If 'everyone's "monuments" are everyone else's "documents" and vice versa' (Panofsky 1955, 10), it is remarkable that in this forgotten *record* of Trikoupis, in the historical center of Athens, new *recordings* call for Trikoupis to be unveiled and seen. As a result, the monument, as a material document that encapsulates the dominant memory of public history, is again relocated, this time symbolically from the *lieux de mémoire*, the spaces of memory, to the *milieux de mémoire* (Rothberg 2010), the place where memory is generated.

Today, the social body of the city, protesting against the modern reality of the Greek crisis in an effort to express its social desiderata and hopes for the future finds in Trikoupis' statue[7] the means to comment on a symbolic level the dominant memory of public history incorporated in the monument as a material document.

Notes

1   Less than a year from his death and just before the Greek-Turkish war broke out, Trikoupis' name entered the Greek pantheon as his public works and reforms were considered crucial in addressing the challenged of the time. ' . . . the whole nation will kneel before the memory of Charilaos Trikoupis and will set up in the capital of the

Hellenic Nation a marble statue of the Great Man' (Anonymous, 'Peri tes Mnemes tou' (On his Memory), *Empros*, A' 86, 05.02.1897, 1. Available at: http://efimeris.nlg.gr/ns/ pdfwin_ftr.asp?c=108&pageid=-1&id=340&s=0&STEMTYPE=0&STEM_WORD_ PHONETIC_IDS=AAiARwASZASRASSASXASlASYASPASaAAgASUASVASPAS UASNASHASXAAi&CropPDF=0 (Accessed 24.05.2019).

2 King's College London project: *The Art of Making in Antiquity, Stoneworking in the Roman World*. Available at: www.artofmaking.ac.uk/explore/materials/8/Pentelic-Mar ble (Accessed 20.05.2019).

3 The placement of the sculptures was the responsibility of the City of Athens, with the final approval of the Prefecture. Their placement usually led to objections among municipal councilors and the Prefecture and press criticism. Respected, however, was the view of the works' creators.

4 Special thanks to Filippos Mazarakis-Ainian, Curator of the National Historical Museum, for bringing the note to my attention.

5 President of the Commission was the architect and archaeologist Anastasios Orlandos (1887–1979). See Union List of Artist Names® Online. Available at: www.getty.edu/ vow/ULANServlet?english=Y&find=Anastasios+Orlandos&role=&page=1&nation= (Accessed 22.05.2019).

6 In the years following the Greek civil war, a tendency to illustrate Greek newspapers with drawings and prints whose visual symbolism was related to the Greek War of Independence becomes quite obvious. Greek modern artists create images —usually appearing in the front pages of newspapers— that reinforce national pride and unity. It is interesting to note that the historical event acts as a common source of symbols for both right-wing (*Akropolis, E Vradyne, Ethniki Efemeris, Ethnikos Kyrex, Empros*) and left-wing (*Demokratikos Typos, Proodeftikos Fileleftheros, Rizos tis Defteras*) newspapers.

7 Members of the Party of Ecologists Greens protest against the closure of much of the country's rail network — Trikoupis was the founder of the railway network in Greece — on June 5, 2020, World Environment Day. Available at: www.youtube.com/ watch?v=xflLYMwS2Xs & www.youtube.com/watch?v=Ejk6zUFmuvM (Accessed 22.05.2019).

Bibliography

Anisios, N. 1914. 'O Andrias is ton Trikoupen (The Statue in Honor of Trikoupes).' In *Empros*, 22.10.1914, 3.

Anonymous. 1912. 'Grammata kai Technai' (Letters and Arts).' In *Pinakotheke*, IB 141(11), 177.

Anonymous. 1920a. 'Ta Apokalypteria tou Andriantos tou Charilaou Trikoupi.Yperochos Logos tou k. Repoule (The Unveiling of Trikoupis' Statue. Mr. Repoulis' Wonderful Speech).' In *Aster* 7766, 27.01.1920, 1–2.

Anonymous. 1920b. 'Ta Apokalypteria tou Andriantos tou Trikoupe, he Chthesine Telete. O Logos tou k. Repoule (The Unveiling of Trikoupis' Statue, Yesterday's Ceremony. Mr. Repoulis' speech).' In *Empros* 8364, 27.01.1920, 1–3.

Anonymous. 1920c. 'O Charilaos Trikoupes.' In *Athinaike*, A, 345, 27.01.1920, 1.

Anonymous. 1920d. 'Ta Apokalypteria tou Andriantos tou Charilaou Trikoupi (The Unveiling of Trikoupis' Statue).' In *Ethnos*, 26.01.1920.

Anonymous. 1920e. 'Ta Apokalypteria tou Andrianta tou Ch. Trikoupi. He Chtesyni Telete eis ten Voulen (The Unveiling of Trikoupis' Statue. Yesterday's Ceremony in the Parliament).' In *Patris*, 27.1.1920.

Anonymous. 1920f. 'Charilaos Trikoupis.' In *Patris*, 26.01.1920.

Antonopoulou, Z. 2003. *Ta Glypta tes Athenas, Ypaithria Glyptike 1834–2004 (Athens' Sculptures, Outdoor Sculpture 1834–2004)*. Athens: Ekdoseis Potamos.

Boime, A. 1998. *The Unveiling of the National Icons: A Plea for Patriotic Iconoclasm in a Nationalist Era*. Cambridge: Cambridge University Press.

Chatziiosef, C. ed. n.d. *Istoria tis Elladas tou 20ou Aiona (The History of Greece in the 20th Century)*, vols A1–A2: *Oi Aparches (The Beginning), 1900–1922*, vols B1–B2: *Mesopolemos (Inter-war period), 1922–1940*. Athens: Vivliorama.

Christou, C. and Koumvakali-Anastasiade, M. 1982. *Neohellenike Glyptike 1800–1940*. Athens: Commercial Bank of Greece.

Clogg, R. 2014. *A Concise History of Greece*, 3rd ed. Cambridge: Cambridge University Press.

Confino, A. 1997. 'Collective Memory and Cultural History: Problems of Method.' In *The American Historical Review* 102(5), 1386–1403. Available at: www.jstor.org/stable/2171069 (Accessed 19.05.2019).

Daniel, M. 2012. *He Merimna tou Demou Athenaion gia ta Demosia Glypta tes Poles 1835–1940) (The Concern of Athens Municipality for the Public Sculptures of the City 1835–1940)*. Available at: www.archaiologia.gr/blog/2012/01/23/η-μέριμνα-του-δήμου-αθηναίων-για-τα-δη/. (Accessed 20.05.2019).

Fessa-Emmanuel, E. ed. 2009. *Ellenike Architektonike Etairia, Architektones tou 20ou Aiona (Hellenic Architectural Society, Architects of the 20th Century)*. Athens: Potamos.

Fowler, D.D. 2017. 'Uses of the Past: Archaeology in the Service of the State.' In *American Antiquity* 52(2), 229–248.

Galinike, S. 2015. *He Chresis tes Archaiotetas ste Sygkrotese tes Taftotetas mias Poles. He Periptose ton Sychronon Glypton tes Thessalonikis (The Uses of Antiquity in Constructing the Identity of a City. The Case of Thessaloniki's Contemporary Sculptures)*. Unpublished doctoral thesis. Athens: School of Architecture, National Technical University of Athens. Available at: http://thesis.ekt.gr/thesisBookReader/id/36255#page/304/mode/2up (Accessed 20.05.2019).

Gell, A. 2013. 'Technologia kai Magema (Technology and Magic).' In Rikou, E. ed., *Anthropologia ke Syghrone Techne (Anthropology and Contemporary Art)*, 1st ed. Athens: Alexandria, 203–232.

Geronta, D. 1972. *Historia Dimou Athinaion, 1835–1971 (History of the Athens Municipality, 1835–1971)*. Athens: Αθηναίων Δήμος.

Halbwachs, M. 1950. *La Mémoire Collective*. Paris: Presses Universitaires de France.

Halbwachs, M. 1992. *On Collective Memory*. Chicago and London: University of Chicago Press.

Hellenic Literary and Historical Archive, National Bank Cultural Foundation. *Photographic Archives*. Available at: http://archives.elia.org.gr:8080/LSelia/images_View/04.15.2.55.JPG (Accessed 22.05.2019).

Hellenic Literary and Historical Archive, National Bank Cultural Foundation, Thomas Thomopoulos' Archive (1884–1967).

Hellenic Literary and Historical Archive, National Bank Cultural Foundation, Thomas Thomopoulos' Archive Digitized Material. Available at: https://bit.ly/2ZdUBvB (Accessed 19.05.2019).

Historical and Ethnological Society of Greece. Available at: www.nhmuseum.gr/en/about-us/historical-and-ethnological-society-of-greece/ (Accessed 20.05.2019).

Ho Attikos. 1954a. 'Attika Emeronyktia. Hoi dyo Siopeloi (Attic Nights and Days. The Two Silent Ones).' In *Empros* 4, 09.01.1954, 2. Available at: http://efimeris.nlg.gr/ns/pdfwin_ftr.

asp?c=108&pageid=-1&id=64926&s=0&STEMTYPE=1&STEM_WORD_PHONETIC_
IDS=ARwASZASRASSASXASlASYASPASa&CropPDF=0 (Accessed 24.05.2019).

Ho Attikos. 1954b. 'Attika Emeronyktia. Vanafsotetes, o Geros (Cruelties, the Old Man).'
In *Empros* 8, 06.02.1954, 2. Available at: http://efimeris.nlg.gr/ns/pdfwin_ftr.asp?c=1
08&pageid=1&id=65060&s=0&STEMTYPE=1&STEM_WORD_PHONETIC_IDS
=ARmASXASTASXASSASXAScASZASmASVASPASa&CropPDF=0 (Accessed
19.05.2019).

Ho Attikos. 1954c. 'Attika Emeroniktia. Ochi Misa Pragmata (Not Half-finished
Things).' *Empros* 9, 13.02.1954, 2. Available at: http://efimeris.nlg.gr/ns/pdfwin_ftr.
asp?c=108&pageid=-1&id=65094&s=0&STEMTYPE=1&STEM_WORD_PHO-
NETIC_IDS=AAiARwASZASRASSASXASlASYASPAAi&CropPDF=1 (Accessed
22.05.2019).

Koliopoulos, J. and Veremis, T. 2010. *Modern Greece: A History since 1821*. Oxford:
Blackwell.

Kouris, T. 2016. 'Paracharaxi e Dikaiose; He Diekdikise tes Mnemes ke tou Chorou Mesa
Apo Ena Palimpsesto Mneimeio (Counterfeiting or Justice? The Claiming of Memory
and of Space Through a Palimpsest Monument).' In *Ethnologhia on-line*, vol. 7, 1–26.04.
Available at: www.societyforethnology.gr/site/pdf/E.o.L.-2016.04-Kouros-%20Paracha
raksi_h_Dikaiosi.pdf (Accessed 22.05.2019).

Lappas, T. 1982. *Ta 100 Chronia tes Historikes-Ethnologikes Etairias kai tou Mouseiou
tes, 1882–1982 (The 100 Years of Historical – Ethnological Society and its Museum,
1882–1982)*. Athens: Historike and Ethnologike Etaireia tes Hellados.

Le Goff, J. 1988. *Histoire et Mémoire*. Paris: Gallimard.

M.C. 1916. 'Kallitechnia. Es to Ergasterion tou Thomopoulou, to Agalma tou Trikoupe
(Artistry. In Thomopoulos' Studio, the Statue of Trikoupis).' In *He Esperia*, A 30,
15–28.07.1916, 476–477.

Matthiopoulos, E. ed. 1997–2000. *Lexiko Hellenon Kallitechnon (Dictionary of Greek Artists)*, 4 vols. Athens: Melissa.

Mykoniates, H. 1996. *Neohellenike Gyiptike (Neohellenic Sculpture)*. Athens: Ekdotike
Athenon.

Nora, P. 1989. 'Between History and Memory: Les Lieux de Mémoire.' In *Representations*
26, 7–25.

Nora, P. 1997. *Les Lieux de Mémoire*. Paris: Gallimard.

Panofsky, E. 1955. *Meaning in the Visual Arts*. New York: Doubleday Anchor Books.

Papademetriou, N. and Anagnostopoulos, A. eds. 2017. *To Parelthon sto Paron. Mneme,
Historia and Archaioteta sti Sygchrone Hellada (The Past in the Present. Memory, History and Antiquity in Contemporary Greece)*. Athens: Ekdoseis Kastaniote.

Rothberg, M. 2010. 'Between Memory and Memory: From Lieux de Mémoire to Noeuds
de Mémoire.' In *Yale French Studies* 118(119), 3–12. Available at: www.jstor.org/sta
ble/41337077 (Accessed 22.05.2019).

Stavridis, S. 1990. *He Symvolike Schese me to Choro (The symbolic Relation to Space)*.
Athens: Kalvos.

Stavridis, S. ed. 2006. 'He Schesi Chorou ke Chronou stE Syillogike Mneme (The Relation
of Space and Time in Collective Memory).' In *Mnimi ke Empeiria tou Chorou (Memory
and Experience of Space)*. Athens: Alexandria.

Stavridis, S. 2010. *Meteoroi Choroi tes Heterotetas (Suspended Spaces of Otherness)*. Athens: Alexandria.

Todorova, M. ed. 2004. *Balcan Identities. Nation and Memory*. London: C. Hurst & Co.

Tomara-Sideri, M. 2016. *Evergetismos kai Neohellenike Pragmatikoteta (Beneficialism and Neohellenic Reality)*. Athens: Economia.

Tricha, L. 2016. *Charilaos Trikoupes, o Politikos tou 'Tis Pteei' kai tou 'Dystychos Eptochefsamen' (Charilaos Trikoupis, the Politician of 'Who is to Blame?' and of 'Unfortunately We Went Bankrupt')*. Athens: Polis.

Verdery, K. 1999. *The Political Lives of Dead Bodies, Reburial and Postsocialist Change*. New York: Columbia University Press.

Voule ton Hellenon. 1986. *Metroo Plerexousion, Gerousiaston kai Vouleuton 1822–1935 (Registry of Plenipotentiaries, Senators and Deputies 1822–1935)*. Greek Parliament: Athens.

Further Reading

Bachelard, G. 1997. *Intuition of the Instant*. Evanston, IL: Northwestern University Press.

Forty, A. and Küchler, S. eds. 1999. *The Art of Forgetting*. Oxford and New York: Berg.

Lydakis, S. 2011. *He Neoellenike Glyptike, Historia – Typologia (Neohellenic Sculpture, History-Typology)*. Athens: Melissa.

12 Dialogues with modernity in the city of Ioannina

Aris Konstantinidis, Natalia Mela, Michael Kanakis, and Paris Prekas

Konstantinos I. Soueref

20th-century Ioannina: beginnings and maturation of modernity

Ioannina, the capital of Epirus, is located directly to the west of the Pindus mountain range, in the basin of ancient Ellopia and Molossia. Archaeological excavations have brought to light the remains of the oldest settlement, within Ioannina's present border, in the area of the castle. The lower part of its walls has been dated to the end of the 4th century BC. The city became known as an important urban center, following the town of Arta, in the period of the Despotate of Epirus in the 13th century. In 1204 took place the Sack of Constantinople by the Crusaders and Michael I Komnenos Doukas, the Despotate's first leader and founder, helped wealthy aristocratic families coming from the great city to flee and find refuge in Ioannina. In 1337, Ioannina became part of the Byzantine Empire and was ruled by governors like Thomas Preljubović and the Tocco family. In 1430 the town surrendered to the Ottomans, from whom it was set free in 1913 during the Balkan Wars. Ioannina reinforced its fortifications and administrative structures in the years 1788–1822, when it was under the rule of the infamous Ali Pasha. Despite the length of the Ottoman occupation, Greek remained the native language of most people and Ioannina became an important Modern Greek Enlightenment center in the 17th and 19th centuries.

Ioannina developed its town planning after 1913. The Castle of Ioannina, which extended across the western and northern peninsula of Lake Pamvotis, with the architectural remains of the Byzantine and Ottoman eras and the settlements in and around its walls had parts that were solely inhabited until 1913 either by the Greek, the Romaniote (Greek) Jews or the Turkish communities; in other words the Christians, the Jews and the Muslims (Papastavros 1998).

Morphological elements, most of them mixed, were added to the buildings in the interwar period, combining local traditions of the Epirote stone workers, with shades of European and neoclassical experiences and of selective modernist initiatives in relation to new materials (Rogkoti 1988; Smiris 2003, 2013). Before the 1950s, the new urban planning gradually expanded outside the walls with the construction of private and public buildings. The architectural vocabulary was picked from an imaginary triangle of Ottoman, Greek, and Western European features. In this context, buildings like 'Hotel Acropol' (1936) of Demosthenes Molfessis, the 'Sanatorium' (1938) of Orestes Maltos, and the 'School of Engineers' of Konstantinos Kitsikis (1948/1950), which introduced the earliest images of modernism, were uneven, timid exceptions of the period.

Substantial aspects of modernism in the architecture of Ioannina appear after 1950, with the 'New Zosimaia School' (1951–1957) and the 'Metropolis House' (1955), both works of Patroklos Karantinos (Giakoumakatos 2009). In particular, Karantinos erected the Bishop's residence next to the church of St. Athanasius, abstractly repeating the motif of the ecclesiastic complex (church and bell tower), with vertical and smaller horizontal and inclined lines in a rather awkward experiment. Among other buildings of the same period can be distinguished the 'Workers' Houses' (1957) of Constantine Doxiadis, 'Palladium Hotel' (1955) of Philip Vokos, and 'Xenia Hotel' (1959) of an architects' group under Philip Vokos and Aris Konstantinidis. All of the aforementioned buildings express the fundamental principles of the Modern Movement.

However, catalytic, new 'arbitrary' entrances in the city center were the branch of the Bank of Greece (1961–1966) of Michael Kanakis and especially the 'Museum of Ioannina' (1963–1966) and the café-restaurant 'Oasis' of Aris Konstantinidis (1971) (Philippides 1997). This now modern cityscape, which enriched the pictorial imaginary of the local society, was subsequently matched by a number of buildings with modernist and innovate elements. The 'Traditional Crafts Center of Ioannina' (KEPAVI) (1989) of Dimitris and Suzanna Antonakaki remains, however, the original complement of the first truly modern architects in Ioannina, Kanakis and Konstantinidis (see the third section below).

In regard to plastic arts in the public urban spaces of Ioannina, until the 1970s monuments were limited to war memorials, busts, and neoclassical statues

Figure 12.1 The National Bank of Greece, architect Michalis Kanakis, (1961–1966).

© Photo and source: Konstantinos I. Soueref.

Figure 12.2 Paris Prekas 'Pyrrhus and Dodona,' 1966.
© Photo and source: Konstantinos I. Soueref.

(Papastavros 2009). Among them can be distinguished the bust of 'Lorenzo Mavilis' (1914) of Peter Roumpos in the homonymous square, and the 'Memorial of Bizanomachon (the Bizani Fighters)' (1936) of V. Falireas in the Park of the Clock.

An important exception is the marble relief of Paris Prekas 'Pyrrhus and Dodona' (1966), which forms part of the façade of the aforementioned local branch of the Bank of Greece (Papastavros 2009). This can be regarded as the most important and innovative modern project in the heart of Ioannina. Its abstract iconography summarizes references to the collective consciousness and sensitivity of Epirus.

The bronze sculpture 'Ioannina, Arms, Money, Letters' (1985) of Kyriakos Rokos at the entrance of the Cultural Center and the 'Heroon' (1986) by Theo Papagiannis in the homonymous square inaugurated a series of modern sculptures, which changed the visual landscape of the city's public spaces, a great deal of which were works of Epirote artists.[1]

The Litharitsia Park

The Square of the 25th of March, or the Barracks, or Lithritsia Park, is located in the public open space east of the central Dodona Road and the City's Town Hall. In the broad hill of the park which overlooks the Lake Pamvotis and the Pindus' Range, Ali Pasha built the summer palaces of his sons, Mukhtar and Veli, which were demolished after 1820, and where later the great Camp was erected. It functioned until the city's liberation in 1913. The conversion of the area into a tree-lined park limited the military complex to the buildings of the 8th Infantry Division. In 1963, the northern part of the park was allocated to the construction of the Archaeological Museum of Ioannina (Konstantinidis 1993; Giakoumakatos 2013; Philippides 2013; Soueref 2013). In the western part of the park, north of the Division, was founded the aforementioned branch of the Bank of Greece. Between 1961 and 1966 the modernist architect Michael Kanakis designed and built it, while the modernist architect Aris Konstantinidis erected the Museum of Ioannina between 1963 and 1966. Also in 1971 Konstantinidis' café-restaurant 'Oasis' was built on Dodona Road, about a hundred meters to the South, in the place of the Ottoman cemetery.

The modernization of Ioannina was particularly defined by the works of Konstantinidis and Kanakis, two important figures of the Greek architectural modernism, who introduced a radical change in the image of Ioannina and the life of its citizens. Dialogues with modernity were launched in three key aspects of the modern city: cultural exchange, the arc of history, and entertainment.

Four landmark projects and their starting points

The focus of attention on the Kanakis–Konstantinidis dipole brings issues of integration of new morphological options to the forefront in regard to the surrounding history and physical space. Konstantinidis, in the Museum of Ioannina, his masterpiece in my opinion, (Figure 12.3), important to the history of Greek architecture as well as that of its museums, uses fair-faced concrete together with

Figure 12.3 Aris Konstantinidis, the Archaeological Museum of Ioannina.

© Photo: Ephorate of Antiquities, Ioannina.

a non-monumental porch and an intentionally axial synthetic arrangement, alternating heights of 3 and 5 meters. The elongated volumes are interrupted by three
patios. The museum, resting in its garden in the north side of Litharitsia Park, has
become part of the topography. The natural light, the eastern aperture to the lake,
and the mountainous perimeter enrich the scene of the building (Philippides 1997,
2013). The local isodomic stone layers in the walls, though not a direct imitation
of traditional morphologies, do constitute an indirect reminder to the tradition of
the Epirote pelekans (stone workers). However, this reminder, in combination
with the connotations from the museum, as well as internal and external functional arrangements places the building in a universal league as a Greek architectural localism.

Kanakis projects the austere, 10-meter-high rectangular marble exterior of the
Bank of Greece, with the impressive trisection of its façade: in the upper half,
a long window is divided through 25 plain vertical columns. Half of the lower
is divided into three parts. The first third corresponds to the main entrance. The
remaining two-thirds is a marble surface that was intended to be decorated with
the relief 'Pyrrhus and Dodona' of sculptor Paris Prekas in 1966 (4.20x0.90 m.)
(Figures 12.1–12.2). The 'Doric' form of the building, with the whiteness of the
marble and the 25 vertical rails of the window's colonnade, and the inscription
'Bank of Greece' in the middle of the façade, clearly state the connection to Greek
identity, which is defined directly. This relief (Figure 12.2) consists of 82 white
marble plaques (Mentzafou-Polyzou 2012). The engravings project highlighted,
giving the illusion of depth. This was a result of sandblasting using steel shav-
ings with brick powder.[2] The choice of the theme came from the historical and
archaeological heritage of the area: Pyrrhus is considered to be the most important
historical figure of Epirus, while Dodona is considered to be the most important
ancient monument close to the city. There are located the sanctuary, the most
ancient oracle of Zeus and Dione, as well as the theatre which has seventeen thousand seats. The Bank of Greece theme is developed in three distinct levels and
includes a multi-figured scene of oracle-giving. It is divided following the gaze
of the rider: down, from the left, the four-horse chariot with the armed Pyrrhus is
followed by eight warriors with shields. In the middle level, Pyrrhus is framed by
four warriors to the left and two to the right side. In the upper part, the Oracle is
articulated by three female figures to the left and four to the right, while the center
is dominated by the *phegos* (the sacred oak tree), the dove, and the oracle-giving
priestess. The relief monumentalizes and displays the Oracle, Pyrrhus and the
Epirotes through these depictions which are schematically and abstractively given
in seven independent episodes.

The content of the project 'Pyrrhus and Dodona' of Paris Prekas is characterized by intense abstraction, as well as by the clear juxtaposition of elements of
the Dodonean cult and King Pyrrhus, unraveling the imaginary, virtual landscape
of the artist's imagination. Taking into account C. Castoriadis's *The Imaginary
Introduction of Society* and J. Lacan, both of whom acknowledge the imaginary
and processes of culmination which exceeds the 'real' and conventional, we could
argue that Prekas presents his own Dodona and Pyrrhus, on the level of images and

emotions, to the passers-by and the citizens of Ioannina. Many of them coincide with him as to feelings toward Pyrrhus and Dodona in their mental landscapes, extending individually, and collectively, the data of History and Archaeology. At the same time, a dialectic develops between Prekas' relief and the 'metropolitan' Museum of Ioannina, in which findings from the archaeological site of Dodona and other regions of Epirus are stored and exhibited in the Hall 'Sotiris Dakaris.' The artistic dialogue of the modernists might be taken as a prelude or an epilogue to the museum, as visitors encounter the façade of the local branch of the Bank of Greece on their way to, or from, the museum.

In 2013 the sculptor Natalia Mela contributed to the artistic dialogue of the modernists Prekas and Konstantinidis by adding the steel 'Hoplites' (Figure 12.4), which overlooks the lake in the eastern museum garden. Mela, Aris Konstantinidis' wife, offered a sound modern epilogue to the contemporary image of Ioannina (Soueref 2014). The growth of the pickaxe, of the shovel and the shield, combined with the line, the circle and the triangle, give life to another imaginative Pyrrhus, as a personification 'of the winner and the loser, of the visionary and the imperious dead.'

These, in my opinion, time-honored, impressive material interventions in the city of Ioannina constitute innovative, pioneering and decisive interventions, a radical change and a transmission of new artistic and architectural episodes in the city of Ioannina, between the 1960s and the beginning of the 21st century, in the

Figure 12.4 Natalia Mela, 'Hoplites,' 2013.

© Photo and source: Konstantinos I. Soueref.

Figure 12.5 Part of the modern sector of Ioannina.

© Photo: Ephorate of Antiquities, Ioannina.

context of the strong local tradition of post-Byzantine, Ottoman and neoclassical morphologies.

Moreover, the works of the pioneers Prekas, Kanakis, and Konstantinidis in the core of the southern expansion of the city center generated and complement the numerous sculptures, many of which of modern and postmodern style, scattered in the wider urban tissue.

Conclusion

To conclude, the modern challenges of the 20th century in Ioannina, both in the center and in other parts of the city, reflect in the best way the inclusion of the city in the modern postwar world, a generalized acceptation of the modernist movement, and a renewed option for the antiquity as a point of reference and as a matrix of ideas and archetypes (Figure 12.5). The component of modernity, on the other hand, as an essential materiality and artistic landscape in Ioannina, can be added to the group of factors which lead to the fair characterization of the city as a multi-cultural and ecumenical universality expressed in the works of Kanakis, Prekas, Konstantinidis, and Mela.

The influential architectural historian Dimitris Philippides (1997) noted some years ago that 'Konstantinidis . . . gave a gift to this city.' My aim in this chapter was to prove that the gifts of modernism to the citizens of Ioannina and their city are more, at least four: the Archaeological Museum of Ioannina, the Branch of the Bank of Greece, the relief 'Pyrrhus and Dodona' in the façade of the bank, and the steel 'Hoplites' in the eastern garden of the museum.

Notes

1 Indicatively, I note: The 'Marble Composition' (1986) of Kostas Dikephalos and 'Peace' (1996) in Litharitsia Park, the 'Monument to National Resistance 1941–1944' (1996) of Evdokia Papageorgiou in the Square of the Cultural Center, the 'Monument of the Epirote Worker' (1996) of Kyriakos Rokos, the 'Monument to the Unknown Master' (2006) by Theo Papagiannis in the Zosimaia Pedagogical Academy, the 'Monument to Ioannina Jews' (1996) by George Chouliaras on C. Averof Street, 'The Couple' (1996) by Nitsa Siniki-Papakosta on Mavili Square, 'The Couple' (1996) by Theo Papagiannis on the Lakeside, 'Tribute to the myth of the Lake' (1996) by Kostas Dikephalos in the park 'Christos Katsaris,' and 'The flight' (2008) by Kostas Kazakos on Pyrrhus' Square.
2 In 1998, works of conservation took place using the encaustic method.

Bibliography

Castoriadis, C. 1999. *L'Institution Imaginaire de la Societé*. Paris: Seuil.

Giakoumakatos, A. 2009. *Stoicheia gia te Neotere Helleneke Architektonike: Patroklos Karantinos (Elements for the Contemporary Greek Architecture: Patroklos Karantinos)*. Athens: Educational Foundation of the National Bank of Greece.

Giakoumakatos, A. 2013. 'O Ares Konstantinidis kai he Epoche tou (Aris Konstantinidis and his Era).' In Soueref, K.I. ed. 2013, 35–54.

Konstantinidis, A. 1993. *Empeiries kai Peristatika (Skills and expertise)*, vol. 2. Athens: Estia, 35–39 (for the Foundation of the Archaeological Museum of Ioannina).

Lacan, J. 2004. *Écrits: A Selection*. New York: W.W. Norton & Company.

Mentzafou-Polyzou, O. 2012. *Paris Prekas*. Athens: Benaki Museum, 162–173.

Papastavros, A. 1998. *Ioanninon Egkomion (Praise for Ioannina)*. Ioannina: Ekdoteke Voreiou Ellados A.E.

Papastavros, A. 2009. *Ioanninon Glyptotheke. Enas Agnostos Kosmos (The Sculptures of Ioannina. An Unknown World)*. Ioannina: Apeirotan.

Philippides, D. 1997. *Pente Dokimia gia ton Are Konstantinide (Five Essays for Aris Konstantinidis)*. Athens: Libro, 93–112.

Philippides, D. 2013. 'Me Peisma kai Pathos (With Stubbornness and Passion).' In Soueref, K.I. ed. 2013, 25–33.

Rogkoti, M. 1988. 'Giannena.' In *Greek Traditional Architecture*, vol. 6. Athens: Melissa.

Smiris, G. 2003. 'Poleodomia kai Architektonike sta Ioannina Prin kai Meta tin Apeleftherose (Urbanism and Architecture in Ioannina Before and After the Liberation).' In *Zosimades* 3, 645–670.

Smiris, G. 2013. 'He Archtektonike ton Ioanninon 1913–2013 (The Architecture of Ioannina 1913–2013).' In Soueref, K.I. ed. 2013, 15–16.

Soueref, K.I. 2013. 'Ektos Chronou, entos Orion. Ares Konstantinidis: O Architektonas tou Mouseiou Ioanninon (Beyond Time, inside Limits. Ares Konstantinidis: The Architect of the Archaeological Museum of Ioannina).' In Soueref, K.I. ed. *Ektos Chronou, entos Orion. Ares Konstantinidis: O Architektonas tou Mouseiou Ioanninon* (Beyond Time, inside Limits. Ares Konstantinidis: The Architect of the Archaeological Museum of Ioannina). Ioannina: Ephorate of Antiquities of Ioannina, 9–14.

Soueref, K.I. 2014. To Semadi tes Technes (The Mark of Art).' In Soueref, K.I. ed., *Pyrrhus King Hegetor*. Ioannina: Ephorate of Antiquities of Ioannina, 50–51.

13 Painting versions of the Athenian landscape

Spyros Vassiliou and Yiannis Adamakis

Anina Valkana

Mutability, the ceaseless and blistering change of contemporary societies, constitutes the key feature of modernism. In the present article an attempt will be made to emphasize the way the urban landscape is perceived and transformed in the later modern and postmodern convention, by focusing on two artists that belong to different time periods and aesthetic backgrounds, Spyros Vassiliou and Yiannis Adamakis. The concept of change, absolutely interwoven with the city[1] acquires here a particular meaning, as both artists live in transitional times, during the postwar the first one, and during the current years of financial recession the second, when the physiognomy of the Athenian capital changes dramatically.

Spyros Vassiliou

Vassiliou was born in the Greek countryside, in the small coastal city of Galaxidi, around 1902 or 1903. He spent most of his life in the city of Athens, from 1921 when he enrolled in The School of Fine Arts, until his death in 1985. Resident of the historical city center, enjoying the Acropolis view, he portrayed ceaselessly and in multiple variations the changing urban landscape throughout his life.

Vassiliou's great interest for Athens is not by itself special, since, from the 1821 Greek War of Independence onwards, the Greek capital was found at the center of attention of foreign travellers and Greek artists, all charmed by the mythology of the Acropolis, the ancient ruins and later the Byzantine ones, as well as the neoclassical architecture. They portrayed the city in an idealized way, containing the glory of the past along with a tendency for Europeanization, through the imposition of the neoclassical and the new civil ethics of the Modern Greek state. During the 1930s, and after highly significant historic events had taken place that caused deep turbulence to the Greek society and changes to the demeanor of urban centers – especially the capital's – the issue of 'Greekness,'[2] that is the redefinition of the national identity, takes the lead in theoretical discourse as well as the artistic work of the Generation of the Thirties, to which Vassiliou himself belongs. Within this framework, landscape is located in the center of Hellenocentric views, revolving around the emergence of firm characteristics of the Greek nature and Greek art, its timelessness, and its European connection. Special qualities and morphologies of the natural and urbanized space, like light and the clarity of lines, are especially prominent in island landscapes, like Nikos Hadjikyriakos-Gkikas' *Hydres* (Valkana 2007, 104–113), while in postwar versions of Athens by the same artist, and others as well, city versions that give prominence to its neoclassical character prevail. Yannis Tsarouchis, for example, chooses ideal versions of Piraeus and Athens which connect the natural background, the quality of neoclassicism, the cultural elements of antiquity and the authenticity and noblesse of the folk (Loukaki 2009).

The neoteric, not as morphologic, stylistic expression of modernism, but rather as a thematic choice that emphasizes the transformation of the city, is not introduced in the visual presentation of the urban landscape by the main representatives of the Generation of the Thirties, because, as their texts[3] show, the eradication of the neoclassical and the folk is experienced as death, as a blow to the authentic, the traditional, the ethnic. Most Greek artists, up until the mid-sixties, insist on nostalgic or ideal representations of the city, when Europeans and Americans of the same art movement had already progressed from the positive readings of modernization to the incorporation of negative sub-statements of the civic in dystopic representations of the city.

Contrary to most Greek artists of the same time, Vassiliou sees Athens as a changing contemporary city. He records mainly the postwar transformation of the urban landscape in direct correspondence to the reconstruction connected to the urbanization of the Greek space and the population explosion, mainly in the Greek capital. The issue of 'land-for-apartment exchange system' (antiparochi), one of the main forms of private activity which led to continuous changes in the appearance and the features of the urban landscape, is underlined in many of his paintings. Which is, however, the city that Vassiliou indeed sees? And how does he see it as a neoteric subject?

In his most characteristic paintings of the 1950s and on, the artist views the city from above: most times from his small loft on 6 Webster street and after 1957 and on from his new atelier, built on the terrace of 5 Webster street. It is through this open window or the terrace of his studio that Vassiliou records the changing view of the Acropolis and of the east side of Hymettus Mountain, initially using his camera and following that, his paintbrush. He stands across the city as a spectator, in most cases[4] away and above it, as commentator of its modernity. This viewing of the city from above or from the air, established in modernism as the 'new vision' (Moholy-Nagy 1932), emphasizes a different way of viewing the world, the dynamics and the geometry of the modern city, as well as the acquisition of modern consciousness. It is not by chance that this viewing from above is adopted post-war in many cinematic captures or panoramic, photographic presentations of Athens, with political, tourist or propagandistic interest that display the magnitude of reconstruction and transformation so as to praise the progressive profile of the capital (Papadopoulou 2016, 39–60).

Vassiliou himself writes about the changing postwar Athens:

> Hulking blocks of frigid concrete, colourless and tasteless, ruled on the drawing-board with the hasty pen of progress, are crowding up in place of the old Athenian houses. The rich spread of earthy colours . . . has disappeared,

to be replaced by expressionless white, with here and there the unblended colours of powder paints . . . spread their icy deliberateness over the facades of those monstrosities called apartment blocks.

(Vassiliou 1969, 100–101)

And at another point:

Will Athens ever again be a city with character? . . . But what about down in the streets, with haphazard arcades letting in a tiny breath of air to relieve the suffocating crush of men and machines, with heights towering ever higher until they threaten to hide not just the Parthenon but God himself, where there are no perspectives and no room to breathe? . . . The human yardstick is one that seems to have been consigned to eternal oblivion.

(Vassiliou 1969, 102, 106)

Although in these extracts the painter highlights the measureless, rhythmless, suffocating, ugly face of the contemporary city in direct contrast to the warmth, humane scale, the beauty mainly of the neoclassical and the natural landscape as well, in his artistic versions of Athens, Vassiliou does not consider modernism as a dystopic reality, but rather suggests, in the first place, an open approach of the urban landscape through his framed layouts. That is, he sees Athens as a palimpsest,[5] as a whole of fragments from different eras and thus in the framework of an evolutionary process that identifies with the history of the city itself he depicts without exemptions the multiple historical layers: the archaeological ruins, the classical monuments, the neoclassical buildings, the modern, the newly erected apartment blocks, and in the shadows of the city, the rocky natural background of the Acropolis Rock and Hymettus Mountain that remain indisputably stable in an all-manner, everchanging landscape. Athens, a city open to progress, is acknowledged as a modern, attractive city mainly thanks to this very clean and exact fragmentation.[6] In this sense, Vassiliou in his work *Homage to Parthenis' demolished house II*, 1970–1973 (Figure 13.1), becomes part of an everyday affair which includes his greeting with Konstantinos Parthenis, one of the proponents of modernism in Greek art, next to the half ruined house of the painter. Despite the anachronism that one can see here, since Parthenis' house[7] – characteristic example of modern architecture – was demolished after his death in 1967, this depiction of the demolition by Vassiliou is a strong comment not only on the way the Greek society and the official state perceive modern architecture, but also a wider criticism on the urban planning and general political choices that in many cases destroy memory in the name of unity (Ramoneda 1994, 15).

Vassiliou expands even further the notion and the sighting of the city, incorporating the private space within the public and inversely, appropriating the idea of unity of internal and external space that appears in the intellectual and artistic movements of the first half of the 20th century as an answer to the problematic and confusing picture of subject and identity (Hall et al. 2010, 417). The fluidity

Figure 13.1 Spyros Vassiliou. Homage to Parthenis's demolished house II, 1970–1973, acrylic in canvas, 90 × 115 cm.

Source: © Spyros Vassiliou Archive.

of limits and the redefinition of the relationship between the private and the public holds a central position in the neoteric rational, with Benjamin claiming that the city, despite the changes, should become an interior for its citizens, meaning the adoption of a more personal, one could say more 'private,' stand toward the urban (Guiheux, 1994). In the works of the Greek artist, the interpenetration of the two spaces is achieved either through the open window and the encapsulated paintings with views of the city on the walls or the floor of the interior, or through the surrealistic 'movement' of familiar objects within the city,[8] as the chair, the mirror, the lamp or even full-genre scenes deeply rooted in the consciousness of the people like *The Ash Monday dinner*. These objects liven up the landscape and at the same time consist a source of memory for the history of everyday life. The title *The microcosm of Webster street* that Vassiliou gives to one of his paintings includes the notion of the private, a world smaller and thus more familiar included or contrasted to a bigger and thus stranger one, the big world of the great city. Vassiliou sees the city as an integral part of his microcosm. The carefree, miniature figures inside the public space express the nostalgia to retain the shapes of the early urban life that favor family ties, the bonds within the neighborhood, sociality and generally the humane face of the city.

On the threshold of reality and imagination, Vassiliou's microcosm comprises a suggested sighting of the neoteric city from a different viewpoint which is connected to a catholic life viewing. The artist talks about humans' primitive instinct to 'grasp the vision' of the world and calls for divine grace so as to be able to see the things around him in a contemplative, dreamy, true and loving way and commune them to the rest (Vassiliou 2002, 184, 253–255). It is this feeling of love that allows him to grasp the greatness of nature. It is his altruistic vision, a peculiar relation of the soul with the world, a stand of detachment[9] from the ego, thanks to which he can attribute an aesthetic value to the everyday and humble and transfigure or emphasize, like common, authentic people and craftsmen even the ugly into something beautiful (Vassiliou 1969, 187, 2002, 174–176). Within the wider spirit of the Generation of the Thirties and the companionate relationships that were developed between people of literature and art, we cannot but acknowledge, in the deep metaphysical relationship of the artist with nature, in his sensuous, positive and engrossed angle, but also in the moral sense that he as an artist holds to transcend the vision of the world, the influence of his spiritual bond with poet Angelos Sikelianos.[10] The painter admired in Sikelianos the love, the dedication and the gift to enjoy even his ailment and 'translate the cold medical term, 'encephalic thrombosis' in a poetic outbreak' (Vassiliou 2002, 256–259).

Figure 13.2 Spyros Vassiliou. *Athens in Gold*, 1966, gold and acrylic in canvas, 113 × 145 cm.
Source: © Spyros Vassiliou Archive.

Confronting the dystopias of transforming Athens, Vassiliou does not suggest a utopic vision of leaving and returning to heavenly conditions. He molds personal mythologies of the city accepting the present, time irreversibility and the inevitable development he transcends through the architectural structure of the city. The nostalgic[11] contemplation for a social prototype that leans to more traditional shapes does not reject the present, but it rather expresses the need for redefinition, as a starting point for the future. Oriented toward the pursuit of the beautiful, even in the most modest, he aestheticizes[12] the urban landscape, isolating aesthetic forms which satisfy the eye. Thus, in terms of modernism, the popular and Byzantine tradition, he transforms the anarchy of the modern city into a geometrical construction and the cold white cement walls to warm surfaces brought to life through the contouring or the thick decorative motifs. He even converts the scaffoldings into rhythmic sets, the graceless television aerials to harmonious vertical axes which set the eyes toward the sky and the 'construction debris' into symbols of the humane pulse and soul of the city. Finally, attributing, as the majority of the Generation of the Thirties, authentic value to light, as the main element of Greekness, he transfigures the color of the sky during sunset into a golden background that laurels the urban landscape (Figure 13.2; *Gold Athens*, 1966) and grants it timelessness and spiritual glory exactly as in Byzantine icons.

Yiannis Adamakis

Adamakis was born in Piraeus in 1959 in a family of seamen. The vastness and unsteadiness of the sea, the continuous voyages of family members and the alternating feelings of joy and sadness that accompanied them, the overwhelming presence of ships and the port, are in the core of his artistic creativity. The urban landscape, also a key issue of his work, cannot be seen separated from his childhood experiences and the remembrance of the sea.

In the current times of financial recession and the consecutive transformations in the social net, Athens is located in a different way at the heart of artistic pursuits. Interest in the capital is connected to a new 'banalité,' as the architect Zissis Kotionis (Kotionis 2006, 51) calls the communal belief for the city that features the dynamics, the most alternative, the most multicultural and maybe its most charming demeanor, contrary to the first postwar decades when aversion for the mechanic and unwelcome urban landscape prevailed. Not only activist and artistic collectivities but also independent creators arduously claim, mainly during the last decade, the reappropriation of public space and the reinvention of its identity through 'excavational–archaeological' style, research infiltrations in the recent or remote past of the city, through mappings of hidden and forgotten aspects of the urban landscape but also personal, inner narrations. At the same time, the artists of the 'new figurative painting,' which seem not to fall short in vitality, see contemporary Athens through a more subjective viewpoint, often with sarcasm, humor, daydreaming (Tsiaouskoglou 2014, 86), or through a confessing and self-revealing disposition, as in the case of Adamakis. In group exhibitions of figurative painters focusing on different aspects of the capital, curators often select a fragmentary language that follows the same picture of the city through which is attempted – here too – the investigation of its multiple identities (Kritikou 2009; Vatopoulos 2018).

The subjective and experiential aspect of the city, which is also identified in contemporary artistic practices related to Athens, was developed since the 1970s in the frame of postmodern discourse about the urban phenomenon as a reaction to theoretical schemes of civil sociology that attempted to read mainly regularities and laws that regulate the urban life (Stevenson 2007, 117). Jonathan Raban in his famous work *Soft City* (1974), via which according to David Harvey the idea of postmodernism is introduced, talks about his own city, London, establishing new means to intake the urban landscape (Harvey 1989, 3). The postmodern city becomes above all the field of action and interplay of its citizens and as such is characterized by instability, motion, and fluidity.

Adamakis' urban landscapes are above all fluid places that include, according to Raban (Stevenson 2007, 118), hints, pictures, signs, and within them or through them experienced moments, everyday experiences, dead ends, and fantasies of the artist are revealed. Adamakis himself when talking about his painting in correlation to the contemporary era where insecurity, inconsistency and turbulence (of balance) prevail, he talks about a journey of memories, confessions and ingenious paths, like the ones of Ulysses, in a quest for inner truth (Adamakis 2017). Streets of the urban center, like Panepistemiou Avenue, Patession, Ermou and Tarella streets, Omonoia Square, the underground stations, like Victoria, and bus stations consist spots of his routings. Urban monuments, Byzantine churches, like Kapnikarea, superb neoclassical, anonymous buildings are coming out forcefully through a foggy atmosphere created by the expressionistic writing, the gestural stroke, the drippings and the rich superimposed layers of color; 'the all-encompassing blue,'[13] as dynamic and unsteady as the sea that the artist experienced as a child, along with the fickle reality he is experiencing today. The blazes of the city, the means of public transport, the vehicles and the motorcycles, alternatingly in motion and stopped at red traffic lights and the seemingly set presence of people, treated not as souls but rather as totally embodied, moving subjects in the public space, complete the contemporary urban landscape.

Adamakis, as a contemporary flaneur, wanders around the city, on foot or in a more current version of transportation, by car, in places often familiar and strongly experienced by himself, recording frantically with his camera whatever touches him. If wandering consists an unconscious action in Beaudlaire and a conscious one in Benjamin, of observation, comment and participation in the urban action which aims in the development of thought and the discovery of hints of social meaning and of collective memory (Stevenson 2007, 118), wandering in Adamakis seems to entail more a process of investigation and emergence of the individuality of the city. The artist says: 'I claim my privacy. I feel invisible, light, weightless of the burden of my experiences. I wash off my ideas . . . I wander again in familiar spaces. . . . writing down experiences, writing down pictures, reliving my thoughts' (Adamakis 2014). Going back to de Certeau and the metaphorical interpretation of walking[14] as an action through which the subject does not only experience the city, but also creates it (de Certeau 1988, 97), one could proportionally say that Adamakis' wandering produces his urban landscape: a space experienced, despite, or mainly due to social and psychological situations implicit with life in a contemporary metropolis, like anonymity and alienation, not oppressively or fearfully, but rather as a source of liberation[15] and creativity.

City narration is identified with self-narration. In older versions of Athens (Figure 13.3; *Tree in Athens* 2005) the city as a paradise garden or urban palimpsest is depicted as a space and time denser, with embodied fragments from mnemonic places, images of natural landscape, random civil symbols, and inlaid personal memories in the form of boxed pictures and verbal notes. The miniature supplements that often interpolate Adamakis' painted pictures, not only on his urban landscapes, to enrich or disturb the main narrative, are 'automatic thoughts' which other times refer to – with childish immediacy – real events and other times consist a figment of the imagination that are being transformed to artistic truths, determining the moment (Kazamiakis 2016). In any case, they create a solid, experientially built world which, while it may at first glance express a pop charm,[16] in reality operates in a deeper way, expressing and challenging in inner, sentimental elaborations. Fernando Pessoa's line from his poem *Naval Ode* 'And you, things of the sea, old toys of my dreams! Shape my inner life outside of me! . . . pure inside me . . . like the messy content of a drawer that was shed on the ground!' (De Campos 2003, 76) consists for Adamakis a principal point of reference, and maybe the closest literature parallel[17] to his painting.

In his landscape production through memory,[18] the use of photography plays a decisive role. An 'obsessive' photograph collector, Adamakis uses his camera lens like a notebook, giving it a documental and mnemonic role. According to Barthes, photography bears the distinctiveness to still time, in a 'dramatic and monstrous' way (Barthes 2007, 128), eternally recreating something that only happened for a moment. As 'the presence of an absence' photography condenses reality and truth, death and its transcendence (Barthes 2007, 158). The recurring comebacks in this 'past present' constitute for Adamakis a long-established practice that bears in its core the question of time and memory, and also a simultaneous immersion in his personal past. Wandering in his photographic archive, the painter pursues a rapprochement of the past through a filter of the present and the artistic expression of his ideas, searching each time for the 'punctum,' that is, the stimuli which will 'poke' him and seduce him beyond the frame of a photograph (Barthes 2007, 43). As far as his technique is concerned, many of the photographic effects are embodied in his work so as to intensify the theatrical and dramatic element.

In recent nocturnal scenes (Figure 13.4; *Night in Panepistimiou Street*, 2018), the endoscopic, almost psychoanalytic dimension of his works is amplified mostly due to the polysemic night background. The moonlit city, however, has always been charming for painters, principally because through it they managed to study beyond the natural and the artificial light and capture, due to the multiple plays of light and shadow, either the phantasmagoric view of the modern city, or the darkest, violent and subversive sides of it (Roncayolo 1994). In the explosive, full of reflections works of Adamakis, the streets of Athens and venues of night

Figure 13.3 Yiannis Adamakis. *Tree in Athens*, 2005, acrylic, 130 × 97 cm.
© Photo and source: Yiannis Adamakis.

Figure 13.4 Yiannis Adamakis. *Night at Panepistimiou Street*, 2018, acrylic, 120 × 160 cm.
© Photo and source: Yiannis Adamakis.

life are depicted with colorful strength and almost more abstract. The artist himself says that his initial stimuli for these paintings were provided by Rembrandt's 1642 work *The Night Watch* (Rijksmuseum, Amsterdam), mainly due to the finesse with which the Dutch artist manipulates the light and the focused dazzles in the hands and faces of his protagonists, transfusing depth and internality in his work. However, the study of light but also the nocturnal urban landscape consists for Adamakis the excuse to discover and confess, in pain and agony, his wild instincts, intimate thoughts, phobias, passions, and wrongs, connected to past experiences and ultimately apologize for his own night, the dark side of his life, asking in a way for 'Freudian' catharsis[19] (Valkana 2019a). His suggestive writing is complimented, in this case and in many others, by the line of the poet 'The night also needs a sip of your blood to trumpet its stars'[20] (Ganas 2013).

The ceaseless and long-lived wandering of the artist in the field of literature and mostly poetry operates, as does the wandering in space and in the photographic archive, like an endless tank of imageries and memories that consciously or unconsciously thicken his artistic world, and in this case further signal the urban landscape. The artist discovers himself through poetry and at the same time reveals himself. He grew up reading Kostas Varnalis, Kostas Karyotakis,

Nikos Kavvadias, Konstantinos Kavafis even from his teenage years, and later met Michalis Ganas, Yiannis Varveris, Fernando Pessoa, Yorgos Chronas, Orhan Pamuk, Thomas Mann, Antoine de Saint-Exupéry (Valkana 2019a). Intertextual references are either denoted through suggestiveness, through inclusiveness and through the affective charge of his artistic writing, or are explicitly denoted by the artist. Besides the critical texts and exhibition catalogues where Adamakis' painting is often prefaced or completed by polysemic and compact poetic words, the painter attempts, through social media, an everyday, almost calendar-like, exhibition of his painting work – and thus himself – surrounded by poetic extracts that unravel his poetic universe, and at the same time open up another, intimate route to understand his painting.

Summing up, the aesthetic of the beautiful, in total dependence from the feeling of love, consists Vassiliou's proposal toward the ever-changing modern city. His works, 'a hearty invitation aiming at the most sensitive part of our psyche,'[21] were loved at his time perhaps more than any other artist's and are still loved today, as they taught us to see,[22] but mostly as they taught us to accept and love. On the other hand, in Adamakis, the urban landscape is articulated through memory. The artist as flaneur wanders in reality, but also metaphorically the city, in his archive and in his poetic word in a continuous pursuit that starts within him and that, as far as the artistic writing is concerned, provides an aesthetic and psychic activation. Irrefutably, the altruistic, positive view of Vassiliou and the endoscopic introversion of Adamakis are suggestions of transcendence and personal resistance toward the dystopias of contemporary reality.

Notes

1 The curators of the great 1994 exhibition in Centre Pompidou entitled *La ville, Art, Architecture en Europe 1870–1993* also consider the city as a palimpsest in continuous transformation (Dethier and Guiheux 1994).

2 Concerning the issue of Greekness and aestheticization of the past, see Tziovas (2011, 294).

3 Yannis Tsarouchis comments: 'neoclassicism was relevant to our national anthem, which was also neoclassic. With our state. Our flag' (Skopelitis 2001). Regarding the destruction of neoclassical buildings, '[it was like] my life was torn apart,' 'they were sudden deaths of close friends that you didn't have time to photograph' (Florou 1999, 126).

4 In some cases the painter includes himself in a small scale within the urban landscape either using a camera, underlying in this way the important role photography had in his work, or in front of his easel.

5 Regarding the city as palimpsest, see also Rossi (1991).

6 Bouras relates the beauty of Athens to disorder and anarchy, to the mixing of styles and multiple layers that comprise the urban shape. These continuous mutations of the forms and the seeming randomness and disfigurement are homogenized through an unconscious aesthetic order that binds its design (Bouras 2009).

7 The house was located on Rovertou Gkali Street underneath the Acropolis. By the end of the 1950s the expropriation of the house was decided, but Parthenis refused it. The house was demolished after the artist's death.

8 Vakalo comments: 'Vassiliou made pop-art before pop-art . . . While though pop-art is bitter . . . Vassiliou can see in it life's poetry' (Vassiliou 1969, 155).

9 Papanoutsos also talks about a special attitude of detachment and severance from practical bonds as a prerequisite of aesthetic experience (Papanoutsos 1976, 14). Vassiliou also had the same detachment, as his daughter Drossoula Vassiliou-Elliott testifies (Valkana 2019b).

10 Sikelianos' inscription for Vassiliou: 'To my Spyros, with whom I exchange our spiritual blood with pen and chisel, we became fraternal in life and death' (Vassiliou 1969, 82).

11 For nostalgia as a quest of timeless values and prototypes in the Thirties Generation, see Tziovas (2011, 321). For nostalgia as a positive feeling that sees the future, see Boym (2001).

12 The use of photography also contributes to aestheticization of space. As Benjamin of the Thirties commented, photography aestheticizes whatever it depicts and thus even a deprived environment acquires through the photographic lens aesthetic qualities that shift attention from real space (Skoufias 2003, 87).

13 The line 'the all-encompassing blue' comes from Michalis Ganas' same-titled poem (Ganas 2013), and it was also the title of a recent Adamakis' exhibition at the Historical Archaeological Museum of Hydra in July 2018 (Adamopoulou et al. 2018).

14 Walking as an artistic and political tool consists a common practice of artistic groups and networks as the nomadic architecture network (Tzirtzilaki and Sinopoulou 2018).

15 The German sociologist Georg Simmel attempts, in the beginning of the 20th century, the association of the modern city with freedom. According to Simmel, the distinctions and isolation of the city lead to a more intellectual attitude toward life contrary to the sentimental relationships of the countryside, which simultaneously consists a source of many forms of freedom (Stevenson 2007, 52–55).

16 One could compare some of the paintings of Adamakis with works of Peter Blake, like *On The Balcony*, 1955–1957, Tate Liverpool. Blake, one of Adamakis' favorite painters, belonged to the innovators of the British pop scene.

17 When Adamakis discovered Orhan Pamuk's novel, *The Museum of Innocence* and the same-titled museum with everyday objects Pamouk created in 2012 in Constantinople, he saw his painting in them (Valkana 2019a; Pamuk 2012).

18 Regarding landscape and memory, see among others Stavridis (2005).

19 Catharsis, as a psychotherapeutic process which aims at soul purification, is connected to memory as a form of confession. As such a liberating process, Adamakis perceives the recalling of memory and the expression of sentiments via the artistic action.

20 The line comes from the poetic collection *Black Stones* (1980) and is included in the catalogue of Adamakis' exhibition *Nocturnes* (Adamakis 2019).

21 Angelos Katakouzenos' Review (Vassiliou 1969, 107).

22 Angelos Delivorias' Review (Paraskenio 1996).

Bibliography

Adamakis, Y. 2014. *'Sten Pole' (In the City)*. Thessaloniki: Art Gallery Metamorfosis Available at: www.culturenow.gr/sthn-polh-ekthesh-toy-giannh-adamakh-sthn-aithoysa-texnhs-metamorfosis/ (Accessed 25.05.2019).

Adamakis, Y. 2017. *Light leak*. Chania: TEDx Technical University of Crete [Internet]. Available at: www.youtube.com/watch?v=t-pb0IjZb-Q (Accessed 25.05.2019).

Adamakis, Y. 2019. *Nocturnes*. Athens: Zoumboulakis Galleries.

Adamopoulou, D., Veroucas, A. and Adamakis, Y. 2018. *The All – Encompassing Blue*. Hydra: Historical Archives and Museum of Hydra.

Barthes, R. 2007. *O Fotinos Thalamos: Semeioseis gia tin Photographia (Camera Lucida: Reflections on Photography)*. Athens: Kedros Publishers.

Bouras, K. 2009. *Arxitektonikes Maties (Architectural Vues). Athens_ Revisited* [part 1] [part 2]. Available at: www.greekarchitects.gr/gr/index.php?about=67&search=%CE% BC%CF%80%CE%BF%CF%85%CF%81%CE%B1%CF%82&x=36&y=8 (Accessed 24.05.2019).

Boym, S. 2001. *The Future of Nostalgia*. New York: Basic Books.

De Campos, A. 2003. *I Thalassini Odi kai alla Poiimata (Naval Ode and Other Poems)*. Athens: Exandas.

De Certeau, M. 1988. *The Practice of Every Day Life*. Berkeley and Los Angeles: University of California Press.

Dethier, J. and Guiheux, A. eds. 1994. *La Ville, Art et Architecture en Europe 1870–1993*. Paris: Centre Pompidou.

Florou, E. 1999. *Giannis Tsarouchis. He Zografike kai he Epoche tou (Yannis Tsarouchis. Painting and his Era)*. Athens: Nea Synora and A.A. Livanis.

Ganas, M. 2013. *Poiimata 1978–2012 (Poems)*. Athens: Melani Publications.

Guiheux, A. 1994. 'Tract pour une Ville Contemporaine Somptueuse.' In Dethier, J. and Guiheux, A. eds., *La Ville, Art et Architecture en Europe 1870–1993*. Paris: Centre Pompidou, 18–19.

Hall, S., Held, D. and McGrew, A. 2010. *I Neoterikotita Simera. Oikonomia, Koinonia, Politiki, Politismos (Modernity and its Futures)*. Athens: Savalas editions.

Harvey, D. 1989. *The Condition of Postemodernity: An Enquiry into the Origins of Cultural Change*. Oxford: Basil Blackwell.

Kazamiakis, K. 2016. *'Apo to Ampeli sto Theatro'. O K. Kazamiakis omilei alla kai sinomilei me kallitechnes. Y. Adamakis ('From Vineyard to Theater'. K. Kazamiakis speaks and discusses with artists. Y. Adamakis)*. Athens: Ianos [Internet] Available at: www.youtube.com/watch?v=6TmNH0kX6pg (Accessed 25.05.2019).

Kotionis, Z. 2006. *Pes, pou eine i Athina (Tell me where Athens is)*. Athens: Agra Publications.

Kritikou, I. 2009. *Synevei stin Athina (It happened in Athens)*. Athens: Mikri Arktos.

Loukaki, A. 2009. 'Piraeus: Ideal Versions of the Urban Lanscape on the Painting of Tsarouchis.' In Kalafatis, T. ed., *Piraeus: History and Culture*. Athens: Heptalofos, 214–261.

Moholy-Nagy, L. 1932: *The New Vision. Fundamentals of Bauhaus Design, Painting, Sculpture and Architecture*. New York: Breuer, Warren, Putman Inc.

Pamuk, O. 2012. *The Museum of Innocence*. Available at: https://masumiyetmuzesi-en. myshopio.com (Accessed 24.05.2019).

Papadopoulou, E. 2016. *To Astiko Topio stin Elliniki Metapolemiki Photographia (The Urban Landscape in the Greek Postwar Photography)*. Unpublished doctoral thesis. Thessaloniki: Aristotle University.

Papanoutsos, E.P. 1976. *Aisthitiki (Aesthetics)*. Athens: Ikaros.

Raban, J. 1974. *Soft City: What Cities Do To Us, and How They Change the Way We Live, Think and Feel*. London: Hamish Hamilton.

Ramoneda, J. 1994. 'Qu'est-ce que la ville ?' In Dethier, J. and Guiheux, A. ed., *La Ville, Art et Architecture en Europe 1870–1993*. Paris: Centre Pompidou, 14–15.

Roncayolo, M. 1994. 'Transfigurations Nocturnes de la Ville. L' Empire des Lumieres Artificielles.' In Dethier, J. and Guiheux, A. ed., *La Ville, Art et Architecture en Europe 1870–1993*. Paris: Centre Pompidou, 48–50.

Rossi, A. 1991. *H architektoniki tis Polis (The Architecture of the City)*. Thessaloniki: University Studio Press.

Skopelitis, S.B. 2001. *Neoklassika tis Athinas kai tou Peiraia (Neoclassical Buildings of Athens and Piraeus)*. Athens: Olkos.

Skoufias, M. 2003. 'Ta Metavatika Topia kai i Anagnosi tous stin Photographia (Transitional landscapes and their reading in photography).' In Manolidis, K. ed., *'Oraio, Frichto ki Aperitto Topion!' Anagnoseis kai Prooptikes tou Topiou stin Ellada ('Beautiful, Terrible and Plain Landscape!' Readings and Perspectives of the Landscape in Greece)*. Thessaly: Nissides, 81–93.

Stavridis, S. ed. 2005. *Mnimi kai Empeiria tou Chorou (Memory and Experience of space)*. Athens: Alexandria.

Stevenson, D. 2007. *Poleis kai Astikoi Politismoi (Cities and Urban Cultures)*. Athens: Kritiki.

Tsiaouskoglou, D. 2014. *Astiko Topio & Anthropini Morfi stin Metapolemiki Elliniki Zografiki (Urban Landscape & Human Figure in the Postwar Greek Painting)*. Unpublished master's thesis. Thessaloniki: Aristotle University.

Tziovas, D. 2011. *O Mythos tis Genias toy Trianta. Neoterikoteta, Ellinikoteta kai Politismike Ideologia (The Myth of the Thirties Generation. Modernity, Greekness and Cultural Ideology)*. Athens: Polis Publications.

Tzirtzilaki, E. and Sinopoulou, R. eds. 2018. *Nomadic Architecture. Walking Through Fragile Landscapes*. Athens: Futura.

Valkana, A. 2019a. Unpublished interview in Greek with Yiannis Adamakis.

Valkana, A. 2019b. Unpublished interview in Greek with Drossoula Vassiliou-Elliott.

Valkana, K.C. 2007. *Nikos Hadjikyriakos-Ghikas. Zografiko Ergo (Nikos Hadjikyriakos-Ghikas. Painting)*. Athens: Benaki Museum.

Vassiliou, S. 1969. *Fota kai Skies (Lights and Shadows)*. Athens: S. Vassiliou & F. Frantziskaki.

Vassiliou, S. 1996. 'Spyros Vassiliou 1902 e (or) 1903–1985.' TV show *Paraskinio*. Greek Radio and Television Network 1 (ERT1).

Vassiliou, S. 2002. *Me to Pinelo, to Kalemi kai ti Grafida (With the Brush, the Chisel and the Pen)*. Athens: Kastaniotis editions.

Vatopoulos, N. 2018. *Anaskafe sten Athena (Excavation in Athens)*. Athens: Ena Contemporary and Kaplanon Galleries.

14 The mythical landscape of Andrei Tarkovsky

Notes on the interpretation of cinematic space in *Stalker*

Stavros Alifragkis

Introduction

Soviet filmmaker Andrei Tarkovsky's cinematic landscapes are generated by and at the same time become projections of an inner, emotional geography, whose coordinates integrate, alter, adjust, and reflect the spaces that contain them. This chapter focuses on *Stalker*, a 1979 science-fiction movie, loosely based on *Roadside Picnic* by Arkady Strugatsky and Boris Strugatsky (2000),[1] and the mystical Zone in particular, a post-apocalyptic industrial wasteland, in order to examine the way cinematic narrations inform poietic processes of space and place.

Contrary to what has been proposed in relevant literature and described in the original novel about the ominous and dystopian character of the Zone, it will become evident, by deploying a phenomenological approach based on selected writings by Heidegger (Krell 1993, 139–212, 343–363) and Norberg-Schulz (1979), that Tarkovsky's Zone is not a threatening place but rather a welcoming one, a place one can dwell. The Zone can be construed as a place with specific *character*, however irrational its geography might be. On the basis of Bachelard's 'dialectics of outside and inside' (1994), contrasted against Heidegger's concept of *boundaries* (Krell 1993, 139–212) and Lynch's *imageability* (1960), it will be argued that the Zone's unique architecture and its interrelation with nature determines the true, inner character of the terrain. The Zone exists as a powerful place through its contradictions and the mutual blurring of the boundaries between inside and outside, humanmade and natural, temporal and everlasting. Even though Foucault's notion of *heterotopia* (Faubion 1998, 175–185) was coined for the critical assessment of urban environments, this research broadens its exegetic value to include special spaces, such as the Zone. A juxtaposition of the characteristics of *heterotopia* against the spatial attributes of the Zone is expected to justify this analogy. Here, architecture acts both as civilisation's tangible manifestation and as condenser of long-lasting cultural processes. This dual role of architecture will be discussed within the framework of Sherrard's (1956) and Campbell's (1993, 2002) investigations on the locality/universality of art and myth.

In particular, this chapter addresses the spatial qualities of the Zone first by delineating its outer frontier in section *Threshold spaces: the boundary* and subsequently debating the special nature of this pulsating boundary in section *Salle d'Attente: the nature of the boundary* Section *From localisation to location: in* search of place moves on to discuss the Zone as a meaningful place, while section *The Zone: a dystopia of the future or a place to dwell?* describes its potential for becoming a place to dwell. Finally, section *Dislocated places and heterotopic habitats* investigates the possibility of inhabiting the heterotopia of the Zone.

Myth and landscape: the Zone in *Stalker*

Threshold spaces: the boundary

Campbell, author in the field of comparative mythology, reserves a special place in his analysis of myths and their symbolic function for the mythical figures who belong to two worlds, one temporal and one that opens to eternity (2002, 40–62). He calls them *threshold figures* or *guardians* and assigns them with fixed qualities as well as specific spaces that reflect their unique features (1993, 77–89). A threshold figure belongs to a singular threshold space. Campbell understands these spaces as 'passage from time to eternity' (2002, 40). They are places that facilitate a temporary stay, a short pause, a brief interval, both in space and time, between our world of material wear and a special universe that represents the other. These places manage to maintain a temporal autonomy; they usually defy the laws of physics and exist in a temporal-spatial dimension of their own, while combining elements from both worlds. In *Stalker*, Tarkovsky gradually introduces us to the whimsical landscapes of the Zone through the stalker, played by Alexander Kaidanovsky, the threshold figure of the guide, the scouter, but also the looter in the original novel. Tarkovsky's mythical figure, who personifies doubt and moral torment (Johnson and Petrie 1994, 149), navigates his audiences across the threshold space of the borderline, into the extraordinary landscapes of the Zone and toward the cryptic *room* that supposedly grants all wishes. The stalker frequents the borderline threshold of the Zone, where he meets people who require his specialized services, such as the professor and the author. In contrast, the Zone is the space that supplies the raw material for the myth, which takes shape in seven episodes.[2]

Salle d'Attente:[3] the nature of the boundary

Augé describes the travellers' space, the vast halls of railway stations and airport terminals, the expanses of roads and railroads, the sterilized cabins of ships and airplanes as archetypal *non-places* (1995, 79). Augé's study on the landscapes of supermodernity juxtaposes what he calls the *anthropological place*, a symbolic and concrete space that consists of sets of relations, history and distinct identity; the historical reference plane of the ethnologist (1995, 42–74), to *non-places* that lack the characteristics of *traditional places*. Augé's *non-places* do not integrate earlier places, instead they seclude them. Our worlds consist of *anthropological places* and *non-places*. *Non-places* become meaningful only through frequentation and not actual experience or Heidegger's *dwelling*. The amount, intensity and density of frequentation classifies these places under different categories. These construct the shared identities of passengers, customers, visitors, or tourists. They do not promote relations or distinctive identities, only homogeneity, solitude, and similitude (Augé 1995, 103). Finally, Augé argues that *non-places* are quite the opposite of a utopia; they are places that do exist but do not contain any kind of *organic society* (Augé 1995, 111–112).

Greek filmmaker Theo Angelopoulos has been described as par excellence the cinematic landscape artist of the Greek hinterland. However, his rural filmic landscapes are counterpointed by visually strong cityscapes that mutually redefine their respective meanings and functions in the diegetic universe of Angelopoulos' movies. This universe often involves thematizing the desolation and impasse brought about by varied conditions of contemporary statelessness (expatriate, refugee, immigrant, etc.) and the consequent absence of a strong sense of belonging. His protagonists roam the urban and rural landscapes of Greece (*Voyage to Cythera*, 1984), the Balkans (*Ulysses' Gaze*, 1995) and, more recently, the world (*Trilogy II: The Dust of Time*, 2008), but always fail to reach their ultimate destination. They are usually portrayed against stark and dimly lit waiting rooms, in

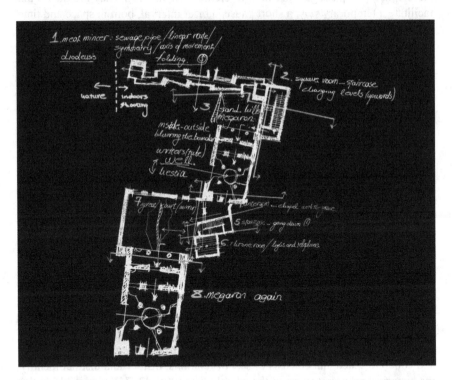

Figure 14.1 Diagrammatic representation of the last three episodes in Tarkovsky's *Stalker*. Consecutive spatial elements of the episodes, the *meat-mincer*, the *sand-room*, the *anteroom* and, finally, the *room* correspond to different spatial layouts from the space syntax of the acropolis of Mycenae (i.e., the corridor, the great courtyard, the megaron).

Source: © Drawing: Stavros Alifragkis.

transit but going nowhere in particular. Sometimes, even entire zones of the city, most noticeably the port of Thessaloniki in *Voyage to Cythera*, function as urban-scale waiting rooms. Angelopoulos' cinematic waiting rooms appear to instantiate very vividly Augé's *non-places* (Alifragkis 2017, 59–63).

One can experience the uneasy co-existence between *anthropological* places and *non-places* in threshold spaces, such as the boundary of the Zone in Tarkovsky's *Stalker*. However, this 'waiting room' experience can be observed elsewhere in the movie, in addition to the borderline *café*, where the stalker, the professor and the writer commence their exploration of the Zone. Their cinematic journey falls short of its original goal; it reaches the *anteroom*, yet another waiting room in the style of Angelopoulos. The fact that Tarkovsky bookends the entirety of his cinematic expedition into the Zone by two waiting rooms – the *café* and the *anteroom* – undermines the physical and geographical underpinnings of the journey and stresses its psychological and emotional aspects.

From localisation to location: in search of place

Heidegger's *dwelling* presupposes that space allows for this *dwelling* to take place. He comments that 'making space for that spaciousness' is elementary in the process of 'setting up a world' (1986, 74–77). These ideas become more concrete in the example of the Greek temple (1986, 68–74). Both its beauty and perfection originate and are reflected in its harmonious coexistence with its natural surroundings and particularly the earth on which the temple lies and, at the same time, emerges from. He notes that 'to make space for means to liberate free space of the open region and to establish it in its structure' and he continues:

> the temple-work, in setting up a world, does not cause the material to disappear, but rather causes it to come forth for the very first time and to come into the open region of the work's world. . . . That into which the work sets itself back and which it causes to come forth in this setting back of itself we called the earth. Earth is that which comes forth and shelters.
>
> (Krell 1993, 171; Heidegger 1986, 77–78)

Heidegger's 'making space for' resonates to a certain extent with Foucault's *emplacement* (Faubion 1998, 175–185). Foucault argues that modern space is a space of *emplacement*, defined by relations of proximity between points and elements, in contrast to the Medieval Ages, where space was hierarchized, thus, forming an ensemble of opposed but interconnected places, a space of localisation (Faubion 1998, 176). In both cases, *space* is used to describe more than a mere three-dimensional organisation of natural elements. Heidegger notes that space:

> made by this manifold is also no longer determined by distances; it is no longer a *spatium*, but now no more than *extensio* – extension. The space provided for in this mathematical manner may be called 'space,' the 'one' space

as such. But in this sense 'the' space, 'space,' contains no spaces and no places. We never find it in any locales.

(Krell 1993, 357; Heidegger 2008, 55–57)

Norberg-Schulz described this *total phenomenon* as *place* and introduced a major distinction between its mathematical aspect, *space*, and its general atmosphere, *character* (Norberg-Schulz 1979, 11–18). Does the Zone constitute a place? Green underlines how the ruins in the Zone create a typically Tarkovskian aesthetic of decay (1993, 93–106). For Turovskaya this sense of dilapidation creates a familiar landscape to those who have witnessed the war (1989, 111). The unfavorable portrayal of the Zone in relevant literature appears to overlook more charitable interpretations of the ruin and its various reconceptualizations in cultural history. Loukaki argues that ancient ruins scattered through bucolic landscapes and their representation in the arts have shaped the way societies envision their relationship to the past and history. These representations constitute idealized landscapes infused with religious symbolism that reflect our conceptions of paradise (2008, 117–121; 2014, 149). Furthermore, one may argue that Tarkovsky's industrial ruins, the silent witnesses of formerly active, humanmade environments transform the landscape from mere space to location. Heidegger is very explicit about this in his example of the bridge:

> [t]he locale is not already there before the bridge is. Before the bridge stands, there are many spots along the stream that can be occupied by something. One of them proves to be a locale, and does so *because of the bridge*. . . . Spaces receive their essential being from locales and not 'space.' . . . They are locales that allow spaces.
>
> (Krell 1993, 354, 355–356, 360; Heidegger 2008, 43–45, 65)

Likewise, the Zone would have not existed as such if it wasn't for the dilapidated carcass of a former factory or a warehouse. In a similar way, the *room* that grants all wishes, the destination of the stalker's journey, is meaningful not only within the environmental and psychological context of its immediate surroundings, but also as a ritualistic procession through space.

When the company begins their journey into the Zone, their guide – the stalker – asks the professor to move toward their first *landmark*, the last utility pole in sight. Lynch's *imageability*, the ability to create mental images that help us navigate through a given environment, appears to be applicable in the case of the Zone and that the stalker is looking for familiar *nodes*, *districts*, and *paths* to do so (Lynch 1960, 46–90). However, the Zone is a place in constant flux, where the organisation of its elements and their interrelation are perpetually reassessed. The fact that the film was shot near Tallinn makes it localisable but not a powerful and distinctive location.[4] Identifying the Zone as a *romantic landscape*, where 'dwelling means to rise up from the micro level' of the lush vegetation 'to the macro level' because of its geographical proximity to Norberg-Schulz's Nordic landscapes constitutes a unidimensional reading of its *genius loci* (1979, 42–45).

Instead, the *character* of the Zone emerges from the sacred unification of both natural and humanmade environments in a oneness that is most strongly experienced along the borderlines, where natural and human creations come together in a mutual awareness.

The Zone: a dystopia of the future or a place to dwell?

Relevant literature describes the Zone in the movie *Stalker* with bias and rather unfavourably; a junk-heap, a dead zone, an oppressive and silent emptiness, a site of catastrophe, a potentially dangerous place, a contaminated area, a human-made environmental catastrophe, an ecological apocalypse. Some even draw parallels between the Zone and the landscapes of the 1986 Chernobyl's nuclear disaster (Tarkovsky 2016, 15 [intr.]; Linaras 2018, 50). Johnson and Petrie adopt a less critical stance by suggesting that the Zone could be both an ominous place and a retreat, acknowledging its admirable qualities and allowing for more positive interpretations (1994, 151–155). This research embarks from this position to propose that the cinematic reconstruction of the Zone illustrates very eloquently the phenomenological notion of *dwelling*. In this respect, the Zone may be construed as something positive, both by virtue of its own spatial traits that qualify it as a place to dwell and in comparison to what lies outside, in the threshold space that surrounds it.

Upon his arrival at the Zone the guide – the stalker – exclaims: 'We are here . . . Home at last!' The stalker expresses his belonging to an environment that he treasures and respects. Here, inside the Zone, he feels free, alive, and complete. Amid the vastness of the natural surroundings, he is secure and confident, with a compelling sense of familiarity that gives meaning to his troubled existence. The stalker, therefore, fulfills the Heideggerian notion of *dwelling* (2008). Heidegger argues that to dwell, to inhabit a place implies 'to be at peace, to be brought to peace, to remain in peace' (Krell 1993, 351) with your surroundings and therefore engage in a passionate relationship with this place in 'the manner which we humans are on the earth, ... [It means] to cherish and protect, to preserve and care for, specifically to till the soil, to cultivate the vine' (Krell 1993, 349; Heidegger 2008, 29). Immediately after his statement, the stalker leaves his company to seek some moments of contemplative solitude in nature. In an almost erotic and mystagogic ritual he reasserts his *primal oneness* with the earth, the sky, and the chthonic deities of the place (Krell 1993, 351). This realisation becomes more evident when the *here* of the Zone is contrasted against the *there* of the world outside the Zone. The opening sequence finds the stalker in his bed, with his family, in a house that cannot be called home. He is merely occupying that space, not inhabiting it. His urge to go back into the Zone is epitomized in yet another statement: 'For me it's prison everywhere!' The term *zone* had been associated with Stalin's labour camps (Johnson and Petrie 1994, 142–143) and, indeed, the stalker both looks like a prisoner and had been imprisoned in the past. Still, the question remains: where does the prison lie? Inside the boundaries of the mysterious Zone or outside?

Tarkovsky's color-coding proves extremely effective in this case. Sequences that take place outside the Zone are shot in sepia, except for two instances, where the stalker's daughter, nicknamed monkey, appears. Researchers unanimously acknowledge that monkey and her bewildering demonstration of telepathic power in the final sequence of the movie suggest hope and victory of faith over the resignation and cynicism demonstrated by the intellectual elite, personified in the professor, played by Nikolay Grinko, and the writer, played by Anatoliy Solonitsyn, who accompany the stalker in his quest. The rest of the film, which takes place inside the Zone, is shot in colour and, in a way, reflects the same optimism. The Zone is painted – literally and symbolically – as mankind's last resort.

Norberg-Schulz points to a crucial aspect of *dwelling*, with reference to Lynch's *imageability* (1960, 9–13, 138–139), which relates to one's inability to get lost: '[t]o be lost is evidently the opposite of the feeling of security which distinguishes dwelling' (1979, 20). The stalker is the only person who can navigate in the wilderness of the Zone. He proceeds with great caution, not out of fear of getting lost but out of respect. His greatest agony is not to disturb the unique equilibrium of the system, though he acknowledges that his very presence affects the status quo and brings about a new balance. The stalker comments: 'It is what we've made it with our condition' and assumes full responsibility for the Zone's temperamental behaviour.

The troubled relationship between humans and nature in the movie *Stalker* recalls Sherrard's criticism of modern societies. Sherrard detects: 'a dynamic relationship in which man's own life is felt to participate in the life of nature, natural phenomena are felt to participate in human experience' (1956, 239).

He comments that, in their individuality humans cannot experience their oneness with everything; they can only think of themselves as separate and independent identities among thousands of other equally separate and independent identities. This casts fresh light to the sad and distressing message of the Zone. The Zone reminds us that we have forgotten to dwell, to be in harmony and peace with our surroundings. We do not inhabit, we abuse, and therefore we cannot experience the oneness with nature and with each other. The Zone is not empty because it is an unwelcoming or threatening place but because we fail to cope outside our material world. The serenity and silence experienced in the Zone are, indeed, oppressive because they reflect this very truth. The Zone is fenced-off and heavily guarded not because it is dangerous, but because it is under threat of extinction. The guide – the stalker – in his naivety comments that: 'Three men cannot spoil the place in one day,' only to be confronted by the professor's cynical reply: 'Why? They can.' If analogies between the Zone and Chernobyl's disaster are possible, these reflect the actuality of the nuclear threat and are not nested in the imagery of Tarkovsky's cinematic landscape.

Dislocated places and heterotopic habitats

Augé points out that the ideal place to contemplate on this world of ours and the extremities caused by the coexistence of *anthropological places* and *non-places* is the deck of a ship putting out to the sea (1995, 89). Foucault comments on *emplacements* that have the unique ability to connect to all other *emplacements* in ways that suspend, neutralise, and reverse the relations designated, reflected, or represented by them. These *emplacements* are of two types: (1) *utopias*, emplacements that have no real place, they represent a society perfected or the reverse of society and they are fundamentally and essentially unreal; and (2) *heterotopias*, real and actual places embedded in the very institution of society. *Heterotopias* are *realized utopias*, localisable but outside all other places (Faubion 1998, 178–179).

The first characteristic Foucault assigns to *heterotopias* is that they do not process geographical co-ordinates. These *emplacements* are not universal but can be classified under two main types: a. *crisis heterotopias*, places reserved for individuals in a state of crisis with respect to society and the human milieu, and b. *heterotopias of deviation*, places destined for individuals whose behaviour deviates from the main or required norm. The fact that the Zone does not recall any specific terrain is apparent throughout Tarkovsky's movie. The Zone is a place, any place, without specific geographical co-ordinates, which gravely concerned censorship in the USSR that was more alarmed with the vagueness and the mystery surrounding its position rather than with the potential anti-Soviet messages of the movie. Furthermore, the Zone seems to function both as a *crisis heterotopia* and as a *heterotopia of deviation*. According to the stalker, the Zone is a place of hope. Those who have lost all hope may appreciate it. The writer is one of them. At the beginning of the movie he admits that he is struggling with writer's block and that he is looking for inspiration in the Zone. However, just before the *room*, he suffers a complete mental breakdown and asks for penance, redemption, and spiritual and ethical resurrection. Furthermore, the stalker is the one whose behaviour deviates from what is considered normal. His wife tells us so in a sequence toward the end of the movie: 'You have probably noticed already that he is not of this world.' The stalker has been rejected from society, probably because of his poor choice of profession, the fact that he is a stalker. He can find console in an environment that poses a threat for others, where only outcasts seek refuge, a place that the rest of the world has intentionally removed from collective memory.

Foucault's *heterotopias* exist in different ways in the course of time. Similarly, the *heterotopia* of the Zone has performed different functions over time: it has been a laboratory – a place of progress and knowledge, a battlefield – a place of horror and destruction, a mysterious landscape – a place of adventurous exploration and a fenced-off area – a place of solitude. Moreover, it assumes different roles at the same time. Exploring this land of mystery and myth can be a joyful and intriguing experience and a dangerous and threatening endeavour at the same time. The Zone is both overwhelming and exhilarating. It assumes different meanings for each member of the group; they all have different reasons for being there, and they expect different things form their journey, but neither the professor nor the writer shares the religious predisposition of the stalker, who experiences their quest as pilgrimage. After all, they represent different ways of reasoning with the world; scientific discovery and intuitive creativity respectively. This is in complete contrast to *non-places*, such as the periphery of the Zone, where the uniform experience is shared by all, without exception or deviation.

A Foucauldian *heterotopia* can juxtapose in a single, real place several *emplacements* that can even be incompatible. This is also true about the Zone. At the beginning of the second part of the movie the company experiences the looping incident that bewilders even the experienced stalker. The Zone is not merely a place that juxtaposes different places but is a *heterotopia* that clones and piles up on itself. Due to the constant fluidity of the Zone each clone-space can be potentially incompatible with the rest, even though all originate from the same spatial DNA. The Zone is a spatial *pli*, a folding space or a temporal whirlpool, whose centripetal forces trigger perpetually renewable, fractal-like reconstructions of its landscapes.

According to Foucault, *heterotopias* relate to temporal discontinuities that he calls *heterochronias*, that is, absolute breaks with traditional time. He files them under two main categories: the first is comprised of *heterotopias of time* that are linked to the accumulation of time and contain all times, all ages, all forms (i.e., museums and galleries); the second consists of festivals or fairs that he calls *chronic heterotopias*, that is, *heterotopias* that are futile and transitory (Faubion 1998, 182–183). Once entering the Zone, the group loses track of time. Space and time expand and distort. Tarkovsky's cinematography, with the mesmerising slow tracking shots, amplifies this sense of prolongment. The decelerated editing rhythm is also affecting our perception of time. This 151-minute-long movie consists of a mere 142 shots (Verhoef 2009). Johnson and Petrie underline Tarkovsky's intention to convey the impression that the whole of the movie was filmed in a single shot, with temporal and spatial discontinuities camouflaged by continuity of action (1994, 152–153). Indicative of this break with traditional time is the sequence where the group rests after what appears to be a long and tiresome day. Yet, it is still daylight and nothing indicates the passing of time. Moreover, the story takes place in an unspecified immediate future that looks simultaneously contemporary and futuristic and therefore timeless. The Zone seems to have always been there, on that earth and under that sky, functioning as a temporal repository, a *heterotopia of time*. Accumulation of time is manifest in the presence of various decaying cinematic props, such as abandoned buildings or rusting machinery; an aesthetic of decay that has contributed to the unfavourable portrayal of the Zone in relevant literature. Here, decay is not a destructive process that leads to death, but a natural process that relates to the circle of life and the temporality of our futile existence. Because of this decay humans may experience the passing of time. This becomes particularly true in the timeless landscape of the Zone, where time rests within every rusty piece of metal, every mouldy and dampened wall, every rotten piece of wood. In their decay lies concentrated and condensed time.

One of the most intriguing features of a *heterotopia* is that it presupposes a porous periphery, a system of openings and barriers that isolates and makes it penetrable at the same time. Foucault comments that 'either one is constrained to enter or submit to rituals of purifications' and adds that others 'look like pure and simple openings but, . . . conceal curious exclusions' (Faubion 1998, 183). Norberg-Schulz (1979, 13, 15) adds:

> [a]ny enclosure is defined by a boundary. . . . The boundaries of a land-scape are structurally similar [to built space] and consist of ground, horizon and sky. . . . The enclosing properties of a boundary are determined by its *openings* . . . . In the boundary . . . character and space come together.

The boundaries of the Zone are explicitly defined by Tarkovsky. The heavily guarded, fenced-off area functions as a porous and organic membrane that isolates and allows for penetration at the same time. A certain amount of interaction, of osmosis between the two worlds is permitted. However, a limit is understood here as the space, where something begins its presence, its *essential unfolding* according to Heidegger (Krell 1993, 352). In this sense, the boundary of the Zone is not the line of the fence but a space with distinct character, an entity between two other entities, whose depth on either side of the borderline cannot be measured precisely. The Zone begins were its presence begins to be felt. This could be the café where the company meets or even the stalker's place. Similarly, what lies outside the Zone begins its presence where the handcar that carries the group inside the Zone stops in the second episode. The space in-between is the place of the limit. This provides essential space for the rituals of purification and initiation described by Foucault to be reconstructed cinematically by Tarkovsky in the sequence with the handcar.

However, the term *outside* has been used in order to describe what exists beyond the presence of the Zone without dispute. Bachelard condemns such oversimplifications when he argues that 'the dialectics of outside and inside is supported by a reinforced geometrism, in which limits are barriers' (1994, 215). The discursive opposition between *outside* and *inside* advances beyond the vastness of outside and the concreteness of inside: '[i]f there exists a border-line surface between such an inside and outside, this surface is painful on both sides' (1994, 217–218). Tarkovsky resolves this binary opposition by reversing and neutralising it. Critics compare the decay *inside* the Zone to an equally crumbling *outside* world or an oppressive and gloomy *inside* to an equally oppressive *outside*. When Tarkovsky constructs cinematically an inside which is similar in formal terms to the outside he does not question the presence of the limit, rather he endows it with new dimensions and asks his audiences to look beyond formal or aesthetic analogies.

The Zone is a constant reminder of the inside-outside relativity. Interior and exterior spaces fuse together in a constant succession of exterior rooms and interior landscapes. The ruins that for Turovskaya evoke memories of the war (1989, 111) seem to constitute a manifestation of the merger between interior and exterior places. Indicative of this mentality is the contrast between the stalker's terrified reaction to the claustrophobic *meat mincer* in the fifth episode and his confidence and serenity when facing the vast exteriors of the Zone. 'The exterior is always an interior' [Le dehors est toujours un dedans] is the title of a section on the wondering gaze of the moving subject from chapter 'Architecture, L'Illusion des Plans' in Le Corbusier's seminal work *Towards a new architecture* (1958, 154–160). The Zone stretches the truth of this observation to test whether the opposite could also apply.

Epilogue: the mythical landscape

Sherrard comments that the break with traditional societies that, for the West, began with the Renaissance, brought about a kind of artist who thinks:

> as a private individual with his own ideas and emotions to express. He is no longer concerned to represent in his work what eternally exists, really and unchangeably, but only what is presented to him by his own immediate and natural environment, or by his memory of it.
>
> (1956, 234)

He attributes this individualism to the collapse of traditional society and its replacement by the humanist culture of the last few centuries. Tarkovsky shares a similar opinion. He prefers the world of spirituality and mysticism of the past over the material world of emotional and intellectual impasse of today. In *The Mirror* (1975) his young protagonist recites a letter by Pushkin that comments on the rich Russian tradition and how it withers away, without being replaced by something equally meaningful. Sherrard adds that the participation of the traditional artist in a living tradition, the experience that comes from within, from one's own subjective world where the eternal ideas are revealed, constitutes the search for supra-individual truth that takes the shape of universal, archetypal ideas and forms (1956, 237).

In the whole of Tarkovsky's filmography there has not been a moment when he does not question his motivation as an artist as well as the sources of his inspiration. He attempted to address not the temporal but the eternal, not the current but the true by virtue of symbolic images. He managed to elevate his consciousness beyond the average, in search of intangible, deeper, inner meanings and attempted to communicate them effectively through his movies. In the movie *Stalker* these meanings take concrete form in the way he manipulates the landscape, the way he fuses the natural with the humanmade, the way he blurs the boundaries between *inside* and *outside* and finally the way he reminds us of the importance of living in harmony with everything around us, the importance that rests in the act of *dwelling*. His places are not idyllic and therefore resist being ideal. However, they surpassed the triviality and temporality of the present and became archetypal in the sense that they refer to any place, any time, anyone, because they address the depths of life itself. It's only then that the work of art stops being individual and becomes universal and timeless. One should seek in these archetypal forms the revelation of their true being and not a formal representation. In this sense, the Zone in the movie *Stalker* becomes less real, less like a place and more like the concept of a place.

Notes

1 This fairly short 'psychological' science-fiction novel by the Strugatsky brothers utilises the story of a brief alien visitation to Earth and the impact the 'leftovers,' as if from an extraterrestrial roadside picnic (Strugatsky and Strugatsky 2000, 102–104), have on the lives of the people living in Harmont, a fictitious town in Canada and one of the six Visitation Zones, as a vehicle for philosophical investigations on faith, culture, science and logic (Strugatsky and Strugatsky 2008, 21–25 [intr.]); typical of the traits that shaped the genre in the wake of the moral crisis brought about by the Vietnam War in the West (Broderick 2007, 48) and as a result of the 'thaw' period in the East (Csicsery-Ronay Jr 2007, 114–115).
2 Due to limitations, the seven episodes, which correspond to equal spatial categories of the Zone, are outlined accordingly: 1. the borderline *café*, 2. the crossing of the boundary, 3. the *graveyard/battlefield*, 4. the *dry tunnel*, 5. the *meat-mincer*, 6. the *sand-room* and, finally, 7. the *anteroom* with its *chapels* (phonebooth etc.) and the *room*.
3 An implicit reference to *Bourani* in the Greek island of Spetses from Fowles' 1965 novel *The Magus*, which functions as the place of myth but also a deception machine (Alifragkis 2016, 164–166).
4 Shooting locations include the outskirts of Tallinn, Jägala River and hydroelectric powerplant, and Iru powerplant (Gamble 2019; Bessmertniy 2014).

Bibliography

Alifragkis, S. 2016. 'Mia Historia Architektonikes Apoplaneses (A Story of Architectural Seduction).' In Pagalos, P. and Alifragkis, S. eds., *Eros – Architektonike – Pole (Eros – Architecture – City)*. Patras: Citylab and PrintUp, 158–172.

Alifragkis, S. 2017. 'Constructing the Urban Cinematic Landscape: Theo Angelopoulos's Thessaloniki.' In Kazakopoulou, T. and Fotiou, M. eds., *Contemporary Greek Film Cultures from 1990 to the Present*. Bern: Peter Lang, 37–69.

Augé, M. 1995. *Non-Places: Introduction to an Anthropology of Supermodernity*. London: Verso.

Bachelard, G. 1994. *The Poetics of Space*. Boston: Beacon Press.

Bessmertniy, S. 2014. *A Unique Perspective on the Making of 'Stalker': The Testimony of a Mechanic Toiling Away under Tarkovsky's Guidance*. Available at: https://cinephili abeyond.org/unique-perspective-making-stalker-testimony-mechanic-toiling-away-tarkovskys-guidance/ (Accessed 10.06.2019).

Broderick, D. 2007. 'New Wave and Backwash: 1960–1980.' In James, E. and Mendlesohn, F. eds., *The Cambridge Companion to Science Fiction*. Cambridge: Cambridge University Press, 48–63.

Campbell, J. 1993. *The Hero with a Thousand Faces*. London: Fontana Press.

Campbell, J. 2002. *The Inner Reaches of Outer Space: Metaphor as Myth and as Religion*. Novato, CA: New World Library.

Csicsery-Ronay Jr, I. 2007. 'Marxist Theory and Science Fiction.' In James, E. and Mendlesohn, F. eds., *The Cambridge Companion to Science Fiction*. Cambridge: Cambridge University Press, 113–124.

Faubion, J.D. ed. 1998. *Michel Foucault: Aesthetics, Method and Epistemology*, vol. II. London: Penguin Books.

Fowles, J. 1968. *The Magus*. London: Pan Books.

Gamble, P. 2019. *Stalker: In Search of Tarkovsky's Soviet Sci-fi Locations* [Online]. Available at: www.bfi.org.uk/news-opinion/news-bfi/features/andrei-tarkovsky-stalker-loca tions (Accessed 10.06.2019).

Green, P. 1993. *Andrei Tarkovsky: The Winding Quest*. London: The Macmillan Press.

Heidegger, M. 1967. *What is a Thing?* South Bend, Indiana: Gateway Editions.

Heidegger, M. 1986. *He Proeleuse tou Ergou Technes (The Origin of the Work of Art)*. Athens and Ioannina: Dodone.

Heidegger, M. 2008. *Ktizein, Katoikein, Skeptesthai (Building, Dwelling, Thinking)*. Athens: Plethron.

Johnson, V.T. and Petrie, G. 1994. *The Films of Andrei Tarkovsky: A Visual Fugue*. Bloomington and Indianapolis: Indiana University Press.

Krell, D.F. ed. 1993. *Martin Heidegger: Basic Writings*. San Francisco: Harper Collins Publishers.

Le Corbusier. 1958. *Vers une Architecture*. Paris: Éditions Vincent Fréal & Cie.

Linaras, Th. 2018. *Stalker: To Megalo Pouthena (Stalker: The Immense Nowhere)*. Thessaloniki: Saixperikon.

Loukaki, A. 2008/2016. *Living Ruins, Value Conflicts*. London: Routledge (first edition 2008: Aldershot: Ashgate).

Loukaki, A. 2014/2016. *The Geographical Unconscious*. London: Routledge (first edition 2014: Aldershot: Ashgate).

Lynch, K. 1960. *The Image of the City*. Cambridge, MA: The MIT Press.

Norberg-Schulz, C. 1979. *Genius Loci: Towards a Phenomenology of Architecture*. New York: Rizzoli.

Sherrard, P. 1956. *The Marble Threshing Floor: Studies in Modern Greek Poetry*. London: Valentine, Mitchell.

Stalker. (1979). [DVD]. Directed by Andrei Tarkovsky. USSR: Mosfilm, Vtoroe Tvorcheskoe Obedinenie [Viewed 10 October 2019].

Strugatsky, A. and Strugatsky, B. 2000. *Roadside Picnic*. London: Victor Gollancz.

Strugatsky, A. and Strugatsky, B. 2008. *Piknik dipla sto Dromo (Stalker) (Picnic by the Road (Stalker))*. Athens: Alpha Omega Ekdoseis.

Tarkovsky, A. 2016. *O Poietes Omoios Theou: He Teleutaia Megale Sunenteuxe (Poet Like God: The Last Major Interview)*. Athens: Manifesto.

The Mirror. (1975). [DVD]. Directed by Andrei Tarkovsky. USSR: Mosfilm [Viewed 10 October 2019].

Trilogy II: The Dust of Time. (2008). [DVD]. Directed by Theo Angelopoulos. Greece: Theo Angelopoulos Film Productions [Viewed 10 October 2019].

Turovskaya, M. 1989. *Tarkovsky: Cinema as Poetry*. London: Faber and Faber.

Ulysses' Gaze. (1995). [DVD]. Directed by Theo Angelopoulos. Greece: Theo Angelopoulos Productions, Greek Film Centre, MEGA Channel, Paradis Film, La Generale d'Images, La Sept Cinema [Viewed 10 October 2019].

Verhoef, R. 2009. *Stalker*. Available at: www.cinemetrics.lv/movie.php?movie_ID=2820 (Accessed 10.06.2019).

Voyage to Cythera. (1984). [DVD]. Directed by Theo Angelopoulos. Greece: Greek Film Centre, Z.D.F., Channel 4, R.A.I., Greek Television, Theo Angelopoulos Productions [Viewed 10 October 2019].

Concluding thoughts

Argyro Loukaki

'Beauty will save the world,' wrote Fyodor Dostoyevsky. But what kind of beauty did he have in mind as capable of this colossal feat? Judging from his oeuvre, the kind that merges with the good and the true in quasi-Socratic manner. During the present critical period where 'all that is solid melts into air,' the weight falls on urban beauty and culture, the urban artistic process and human attachment to consolidate the powers of compassion, action and understanding, but also to open up ways of recovery, experimentation and progress. What preceeded in this volume validates the initial hypothesis that art is the lab and locus for the elaboration of traumas, possibilities, suspended or decommissioned manifestations, as well as of hope and consolation, however bittersweet.

Processes of cultural diffusion, artistic imitation, urban formation, and response to extremely rich cultural palimpsests, plus sanctity, symbolic density but also myth as constituents of the urban space were investigated here around leading cities Athens, Constantinople, Thessaloniki and Jerusalem, smaller cities like Ioannina, many cities globally, or post-apocalyptic urban spaces like the Zone.

Greece is rediscovering the old thread in new ways, highlighting spatialities and possibilities of independent expression and direct democracy. We see this in relevant urban theatrical happenings (Chapter 8) and on urban façades (Chapter 9). Classicism palpitates and breathes on the neoclassical, quaint walls of 'agonistic' neighbourhoods like Exarcheia; this indicates the importance of continuities linking cultural times, especially because they are not conscious or imposed but emerge spontaneously from the grassroots. Ancient and modern intermingle constantly, and not just in official or grand manners or narratives. For instance, just like the Japanese woodcuts, Byzantine icons and wall paintings have a genuinely popular character and purpose, addressing as they do the eyes and soul of spectators in immediate and quasi-democratic manner.[1] Episodes from the life of the Virgin (Chapter 5) resemble sequences of comics. The inherent potential of deeply rooted popular artforms of high pedigree is open to further theoretical exploration and creative praxis alike, based on the 'modernism' of ancient art (Loukaki 2014/2016).

The material aspect of the city and its art, be it marble, paint, mosaic, stone, metal, glass, concrete, or plastic, water surfaces or vegetation does matter a great deal, as does their symbolic aspect as well (Chapters 1 and 2). Destruction, profanation, and desacralisation of the cultural heritage,[2] especially of the first order

(Chapter 1), namely monuments kneaded with the hopes, wishes and blood of a whole people, can have detrimental consequences which poison the deepest layers of the personal and collective self (Chapter 2). This is a fact well attested in the European space throughout its histories, reversal of fortunes, or accidents – take, for instance, Warsaw's World War II destruction, or, recently, the Paris Notre-Dame fire. It is especially so in countries which have undergone long national crises. Because a delicate balance must be accomplished, whereby the expression of wrath and discontent does not dismantle the urban tissue and art which are knit, consciously but also unconsciously, with and around classicism as an everlasting promise of immortality, as utmost Hegelian aesthetics.[3] The sovereign voice of neighbourhoods and communities should be expressed and respected.

It is argued sometimes that culture is the opposite of power's unlovely sway. But culture also reverberates power of various sorts and in various ways, as we saw.[4] Debord's belief in the necessity of cultural rules is still valid; given all this nuanced creative landscape, it is absolutely imperative first, to develop some kind of cultural accord, enunciated or implied, among players of the field; and second, to empower the artistic and cultural dynamics from all social niches during such times, in view of the social vitality they denote. This is easier said than done for various reasons. First, because culture is more susceptible than ever to the workings of global capital, as Adorno and Horkheimer so accurately anticipated early on; we have seen their tight embrace in the case of some particularly gifted street artists. Then, right at this moment, the national and the globalized, post-national condition are at odds; fragilised countries like Greece, the 'usual scapegoat' of the European Union for many years now, appear to be losing this battle not just on the economic but on further fronts: infringements against the urban heritage at large have been unrelenting for far too long in the country's capital city. However, what preceded is emboldening because, despite everything, the arts are reaching staggering moments in crisis-hit Athens and Greece, occasionally outside, or despite, channels of the market.

Granted, insurgences are vitally important fronts of resistance and solidarity against the insatiable arbitrariness of global capitalism (Chapter 7). However, unrelenting insurgences do pose grave issues because they end up attacking the pride, dignity, and myth of a community (Chapter 2). These are not just important locally, but also a steady, authentic bridge to the global.

Yet ironically, in Athens, insurgent art and actions introduce intense contradictions as they both dispute *and* reverberate the classical rhizomes spontaneously and profoundly. Athens continues to 'live its myth' by means of both high and alternative culture (Chapter 9). This signals, and reveals, unstructured channels of cultural continuity and of a circularity which is as paradoxical as it is painful: While some activists hammer or spray paint on neoclassical marble, others imitate classical architecture and sculpture on neoclassical urban façades.

Byzantine cities, real or 'iconic,' tested moral limits, imitation of the ancient and the sacred, the many aspects of the divine, transferral of building models to the periphery, the limits between sanctity and profanity (Chapters 3,4,5,6). The neoclassical space and sculpture (Chapter 11) are important as direct mimesis of classicism and as memory anchorages but also as safe, quiet backdrops to the hectic urban theatricality; this aesthetics of a small-scale, comforting, routine humanity, similar *and* dissimilar to the grandiosity of classicism (Chapter 2), was rendered with reflective melancholy by artists like Tsarouchis and Vassiliou (Chapter13). Classical, but also Byzantine, neoclassical, and modern urban spaces emanate all kinds of archetypal beauty, sublimity, urban constitution, mental accomplishments, and human dignity; the urban stratigraphy of cities is highly important and bonded to be simply erased or abused.

The importance of the city and of urbanity as epicentres of cultural and aesthetic social processes is corroborated here. 'Official' painting now continues to capture such fleeting processes (Chapter 13) alongside graffiti (Chapters 8 and 9). The power of the optical (Freedberg 1989), especially in a tradition like the Greek (Levin ed. 1993), but also of the filmic and mythological (Chapter 14) is indeed immense, expanding and revealing new spatialities. The same is true of the meta-optical in the present technologically defined global landscape (Summers 2003; Loukaki 2014/2016), especially as the consequence of unprecedented global circumstances like the coronavirus epidemic of 2019/2020. Simultaneously, however, other sensorial dimensions of the city, hearing, touch, and smell, portrayed by 'official' painting (Chapter 13), are unexpectedly intensified as parts of the urban experience in the ultramodern cities of high technology, alongside the urge to form personal collectivities (Chapter 8).

The deep desire for the paradisiacal in the urban space is heightened under conditions of crisis. The hunt for ideal utopias continues in all the areas discussed, both spatial and artistic. This involves the political dimensions of the city-garden as a national and democratic imaginary (Chapter 10). The same goes for national distinctions and accomplishments, the importance of which persists despite or independently of the rage of global neoliberalism. This is evident in the *genius loci* of smaller cities like Ioannina and their modernism (Chapter 12).

During critical periods, to capture all such processes which are creative as well as intensely spatial, it is necessary to abolish the traditional closures of art-historical and other perspectives, since the importance of space as a social, aesthetic, economic, geographical, and symbolic field has become acknowledged and deeply sensed.

We need connectors at this difficult juncture, at least as much as we need theoretical, artistic, and agonistic frontrunners; a creative accord is absolutely essential between artists, intellectuals, and societies. I hope that this volume can contribute to this.

Notes

1 For detailed analysis the Byzantine gaze and on Japanese art see respectively Loukaki (2007/2009, 2014/2016).
2 For the sacralisation of Eleusis, a major classical archaeological locus in Attica, and its impacts, see Loukaki forthcoming.
3 See https://plato.stanford.edu/entries/hegel-aesthetics/, accessed 20.05.2020.
4 See, for instance, the foundational role of papacy in Rome's Baroque (Loukaki 2014/2016).

Bibliography

Adorno, T. and Horkheimer, M. 2002. *Dialectic of Enlightenment*. Stanford, CA: Stanford University Press.

Debord, G. 1994. *The Society of the Spectacle*. Cambridge, MA: Zone Books.

Freedberg, D. 1989. *The Power of Images-Studies in the History and Theory of Response*. Chicago: The University of Chicago Press.

Levin, D.M. ed. 1993. *Modernity and the Hegemony of Vision*. Berkeley and Los Angeles: The University of California Press.

Loukaki, A. 2007/2009. *Mesogeiake Politistike Geographia kai Aistheteke tis Anaptyxes: He Periptose tou Rethymnou (Mediterranean Cultural Geography and Aesthetics of Development. The Case of Rethymnon, Crete)*. Athens: Kardamitsa.

Loukaki, A. 2014/2016. *The Geographical Unconscious*. London: Routledge (first edition 2014: Aldershot: Ashgate).

Loukaki, A. Forthcoming. 'Desacralised Eleusis. Contemplating the Impacts of Non-Restoration on a Major Classical and Poetic Locus.' In *Journal of Greek Archaeology*.

Summers, D. 2003. *Real Spaces: World Art History and the Rise of Western Modernism*. London:'Phaidon.

Index

Printed in the United States
by Baker & Taylor Publisher Services